A HANDBOOK OF NUMERICAL MATRIX INVERSION AND SOLUTION OF LINEAR EQUATIONS

A HANDBOOK OF NUMERICAL MATRIX INVERSION AND SOLUTION OF LINEAR EQUATIONS

JOAN R. WESTLAKE

IBM Federal Systems Division
formerly with Control Data Corporation

ROBERT E. KRIEGER PUBLISHING COMPANY
HUNTINGTON, NEW YORK
1975

Original Edition 1968
Reprint 1975

Printed and Published by
ROBERT E. KRIEGER PUBLISHING CO., INC.
645 NEW YORK AVENUE
HUNTINGTON, NEW YORK 11743

Printed in the United States of America

Library of Congress Cataloging in Publication Data

Westlake, Joan R
A handbook of numerical matrix inversion and solution of linear equations.

Reprint of the ed. published by Wiley, New York.
Bibliography: p.
1. Matrix inversion--Handbooks, manuals, etc.
2. Equations, Simultaneous--Handbooks, manuals, etc.
I. Title.
[QA188.W47 1975] 519.4 74-26623
ISBN 0-88275-225-1

PREFACE

A Handbook of Numerical Matrix Inversion and Solution of Linear Equations is intended to be nearly encyclopedic, and was written with the hope that it might provide a single reference source for the "average" scientific programmer with an A. B. and a mathematics major. It attempts to present a straightforward account of a great many numerical methods for the inversion of matrices and solution of systems of linear equations, with emphasis on computer solution. Some of the methods that are not especially useful or practical as digital computer methods are barely touched on, and are included only for completeness, whereas methods that are appropriate for use on digital computers are usually described in greater detail. Few proofs are given, but all should be available in the literature listed in the references. It is assumed that the reader will make extensive use of the glossary, Appendix A, for definitions of unfamiliar or forgotten terminology, and that he will refer to the list of theorems in Appendix B for a summary of basic facts about matrix algebra. Appendix C is intended to provide a variety of test cases as well as methods for generating more. Because the appendices and references are provided as supplementary material, they contain many references, definitions, and theorems not specifically referred to in the text.

The number of references given for those iterative methods related to the solution of partial differential equations is necessarily limited; to do the subject justice would require another book. However, extensive bibliographies for these methods may be found in the references: in Householder [20], Varga [40], and Young [113]. A very brief general discussion of the process by which a linear system of equations arises from an elliptic partial differential equation is presented in Section 3.3.2.1.

It is suggested that the reader who is attempting to select a suitable method for solving a particular linear system should first read Chapter 1, Introduction; Section 3.1, General Iterative Procedures; and Chapter 8, Comments and Comparisons.

This handbook was first written as an internal report for Control Data Corporation, Los Angeles, California, who sponsored most of the revisions necessary for its publication in book form, and provided computer time.

Special thanks are given to Camilla Muriby, who programmed many of the methods described in the manuscript, and whose comments, suggestions, questions, and corrections were extremely valuable in the revision of the original version. Thanks are also due to Naomi Collmar, to Marcia Francis, and, especially, to Helen Porter for their excellent and careful typing of a very difficult manuscript.

Joan R. Westlake

Pacific Palisades, California
January 1968

CONTENTS

A HANDBOOK OF NUMERICAL MATRIX INVERSION AND SOLUTION OF LINEAR EQUATIONS

1

INTRODUCTION

The problem under consideration is that of inverting an $n \times n$ matrix, A, numerically, or of solving the r systems of linear equations $AX = B$, where B and X are matrices of n rows and r columns. The two problems are equivalent in the sense that, given A^{-1}, the solution to $AX = B$ is $X = A^{-1}B$, whereas the solution of $AX = B = I$, with B equal to the $n \times n$ identity matrix, gives $X = A^{-1}$. It will be shown in Chapter 7, however, that A^{-1} should be obtained only if it is needed explicitly, never merely to find $A^{-1}B$. There are a few methods intended solely for finding A^{-1}; most methods consider the general problem: solve $AX = B$. If $r = 1$, write $Ax = b$. (In general the elements of A will be assumed to be real numbers, although in some cases complex matrices are considered.)

The literature on these methods (both classical and current) is so vast that no survey could hope to be exhaustive. A tentative classification and bibliography on solving systems of linear equations, which was published by G. E. Forsythe in 1953 [61] contains over 400 titles!

The numerical procedures to be considered may generally be divided into two classes—the direct methods and the indirect (or iterative) ones. Direct methods would yield an exact solution in a finite number of operations if there were no roundoff error. For the actual approximate solution that is obtained the number of operations performed can be specified in advance, so it is independent of the accuracy desired. Iterative processes, on the other hand, begin with an approximate solution and obtain an improved solution with each step of the iteration, but would require an infinite number of steps to obtain an exact solution in the absence of roundoff error. The accuracy of the solution does depend on the number of iterations performed.

A special case, which could fall into either category, is the conjugate gradient method, a semi-iterative procedure that should theoretically, in the absence of roundoff error, yield the exact solution in at most n steps, but is usually iterated further until the desired accuracy is obtained.

Direct and iterative methods are sometimes combined. For example, a solution obtained by a direct method can be improved by the iteration suggested in Section 3.4.3. Also, in some of the block iterative procedures, subsets of the equations are solved by direct procedures within the course of an iteration.

A comparative discussion of direct versus iterative methods is given in Chapter 8, where an attempt is made to cite situations appropriate for the use of each. Various direct methods are then compared and numerical results as well as some recommendations are supplied. A similar discussion for various iterative methods includes tables of storage requirements and convergence properties. Section 3.1 is a prerequisite for the understanding of most of the iterative processes, since basic definitions as well as further subclassifications of iterative methods are given there.

Brief accounts of acceleration procedures, ill conditioning, error measures, and scaling have also been included.

2

DIRECT METHODS

2.1 CRAMER'S RULE [19, 21]

This method is too well known to elaborate on and too cumbersome for practical use (except possibly for 2×2 matrices), so that only the result is given. If $Ax = b$, then $x_i = |A_i|/|A|$, where A_i is the matrix obtained from A by replacing its ith column by the vector b. Equivalently,

$$A^{-1} = |A|^{-1}(A_{ji}) = |A|^{-1} \quad (\text{adjoint } A).$$

Hence the solution of a system of n linear equations by use of Cramer's rule requires the evaluation of $(n + 1)$ determinants of order n. If evaluated directly, each determinant involves $\alpha n!$ multiplications, where $1 \leq \alpha < e - 1$. Solution of the entire system then uses $\alpha(n + 1)!$ multiplications, n divisions and $(n + 1)!$ additions or subtractions. This is the equivalent of about $2(n + 1)!$ multiplications.

The number of operations needed to evaluate a determinant can be reduced by first reducing the matrix to triangular form by a Gaussian elimination procedure (see Section 2.3.3). The value of the determinant of this triangular matrix is just the product of the diagonal elements times $(-1)^k$, where k is the number of row interchanges which were performed in the reduction to triangular form. This method of determinant evaluation is attributed to Chio and involves $(n - 1)(2n^2 - n + 6)/6$ multiplications, $\frac{1}{2}n(n - 1)$ divisions, and $\frac{1}{6}(n + 1)(2n + 1)n$ additions and subtractions. The solution of n linear equations using Cramer's rule with Chio's method of determinant evaluation thus takes about $\frac{1}{3}n^4$ multiplications.

2.2 METHOD BASED ON THE CAYLEY–HAMILTON THEOREM [8, 27]

Let

$$P(x) = x^n + \sum_{j=1}^{n} p_j x^{n-j} = 0$$

be the characteristic equation of the matrix A, that is, $P(x) = |A - xI| = 0$. By the Cayley–Hamilton theorem (Appendix B, Theorem 23) we have $P(A) = 0$. Hence

$$A^n + p_1 A^{n-1} + p_2 A^{n-2} + \cdots + p_{n-1} A + p_n I = 0.$$

Solve this equation for I, then multiply both sides by A^{-1} to obtain

$$A^{-1} = -(p_n)^{-1}[A^{n-1} + p_1 A^{n-2} + \cdots + p_{n-1} I].$$

If $S_r = \text{Tr}\, A^r$, that is, S_r is the sum of the diagonal elements of A^r, then it is known that

$$p_i = -\frac{1}{i}\left(\sum_{j=1}^{i} p_{i-j} S_j\right)$$

for $i = 1, \ldots, n$, where $p_0 = 1$ (See Appendix B, Theorem 68). Once the p_i have been computed, A^{-1} can be found. The method is not particularly practical because powers of the matrix A must be formed and stored, and of the order of n^4 multiplications are needed.

2.3 METHODS BASED ON TRIANGULARIZATION

2.3.1 General Theory [7, 8, 11, 19, 27, 59, 60, 63, 84, 86]

Procedures based on triangularization as well as those based on diagonalization are probably the most generally useful because they apply to all types of matrices, are easy to program, and, according to many users, seem to rate highly in comparative tests with respect to both speed and accuracy. Every program library should contain at least one of these methods.

All these processes are based on the fact that any matrix A is equivalent to a triangular matrix, and this triangular form is obtained from A by a finite number of elementary operations to the rows and/or columns of A. If A is real, then the resulting triangular matrix is also real.

An equivalent statement, for a nonsingular matrix A, is that all these triangularization procedures are based on the LDU theorem (Appendix B, Theorem 1), which states that A may be expressed (or rearranged so that it may be expressed) as $A = LDU$ (L lower triangular, D diagonal, and U upper triangular), provided that A is nonsingular. The only real difference in the methods lies in the way D is partitioned. D may be incorporated wholly or partly with L or U. D may also be partitioned so that the diagonal terms of L and U are equal in absolute value.

Suppose $A = LDU$ is written as given below, without specifying whether D belongs to L, U, or both.

$$\begin{pmatrix} l_{11} & 0 & 0 & \cdots & 0 \\ l_{21} & l_{22} & 0 & \cdots & 0 \\ \cdot & \cdot & \cdot & & \cdot \\ \cdot & \cdot & \cdot & & \cdot \\ \cdot & \cdot & \cdot & & \cdot \\ l_{n1} & l_{n2} & \cdot & \cdots & l_{nn} \end{pmatrix} \begin{pmatrix} u_{11} & u_{12} & \cdots & u_{1n} \\ 0 & u_{22} & \cdots & u_{2n} \\ \cdot & \cdot & & \cdot \\ \cdot & \cdot & & \cdot \\ \cdot & \cdot & & \cdot \\ 0 & 0 & \cdots & u_{nn} \end{pmatrix} = \begin{pmatrix} a_{11} & \cdots & a_{1n} \\ \cdot & & \cdot \\ \cdot & & \cdot \\ \cdot & & \cdot \\ \cdot & & \cdot \\ a_{n1} & \cdots & a_{nn} \end{pmatrix}$$

We wish to solve the equations which result for l_{ij}, $i > j$ and u_{ij}, $i < j$ with a_{ij} known. After multiplying the two matrices L and U and equating the result to the corresponding element of A, the equations may be written

$$\sum_{k=1}^{\min(i,j)} l_{ik}u_{kj} = a_{ij}$$

or

$$\begin{aligned} l_{11}u_{11} &= a_{11} \\ l_{11}u_{12} &= a_{12} \\ &\vdots \\ l_{11}u_{1n} &= a_{1n} \end{aligned}$$

$$\begin{aligned} l_{21}u_{11} &= a_{21} \\ l_{21}u_{12} + l_{22}u_{22} &= a_{22} \\ l_{21}u_{13} + l_{22}u_{23} &= a_{23} \\ &\vdots \\ l_{21}u_{1n} + l_{22}u_{2n} &= a_{2n} \end{aligned} \qquad \begin{aligned} l_{r1}u_{11} &= a_{r1} \\ l_{r1}u_{12} + l_{r2}u_{22} &= a_{r2} \\ &\vdots \\ l_{r1}u_{1r} + l_{r2}u_{2r} + \cdots + l_{rr}u_{rr} &= a_{rr} \\ &\vdots \\ l_{r1}u_{1n} + l_{r2}u_{2n} + \cdots + l_{rr}u_{rn} &= a_{rn}. \end{aligned}$$

Since there are $n^2 + n$ unknowns and only n^2 equations, n of the unknowns may be specified. In some of the methods that follow, the diagonal elements of L or of U will be specified as all ones; in others, the diagonal elements of L and U will be assumed equal.

Some of the methods considered depend explicitly on the triangular resolution in the sense that all elements of L and U are computed and displayed. In other methods, one of L, U is not explicitly computed.

Wilkinson [41] has shown that the main error arising in the solution of $AX = B$ by triangularization is that resulting from the decomposition of A to LDU rather than from the solution of the two triangular systems obtained.

2.3.2 Triangular Systems

It will be convenient for later reference to write down the solutions of both upper and lower triangular systems.

The system $UX = B$, where U is upper triangular, has the solution

$$x_{ij} = \frac{b_{ij} - \sum_{k=i+1}^{n} u_{ik}x_{kj}}{u_{ii}}$$

for $j = 1, \ldots, r$, and $i = n, \ldots, 1$.

In case $B = I$, $U^{-1} = (x_{ij})$ is obtained, where $x_{ii} = 1/u_{ii}$ for $i = n, \ldots, 1$ and

$$x_{ij} = \frac{1}{u_{ii}}\left(-\sum_{k=i+1}^{j} u_{ik}x_{kj}\right)$$

for $i = n - 1, \ldots, 1$ and $j = n, \ldots, i + 1$.

The system $LX = B$, where L is lower triangular, has the solution

$$x_{ij} = \frac{1}{l_{ii}}\left(b_{ij} - \sum_{k=1}^{i-1} l_{ik}x_{kj}\right)$$

for $j = 1, \ldots, r$ and $i = 1, \ldots, n$.

When $B = I$, the result is $L^{-1} = (x_{ij})$ where $x_{ii} = 1/l_{ii}$ for $i = 1, \ldots, n$ and

$$x_{ij} = \frac{1}{l_{ii}}\left(-\sum_{k=j}^{i-1} l_{ik}x_{kj}\right)$$

for $j = 1, \ldots, i - 1$ and $i = 2, \ldots, n$.

2.3.3 Gaussian Elimination (Pivotal Condensation)

2.3.3.1 Method [8, 11, 13, 19, 22, 41, 42, 59, 63, 108]

Elementary row operations are performed on A (and on B) to annihilate successive elements of A in such a way as to reduce A to upper triangular form U. The product of the matrices effecting the elementary row operations on A is a lower triangular matrix L with unit diagonal. Then $LA = U$ or $A = L^{-1}U$. Since L^{-1} is again lower triangular, A has been expressed in LDU form with D included in U. U is found explicitly, L or L^{-1} is not.

The process is as follows. At the kth stage, a pivotal element $a_{ij}^{(k-1)} \neq 0$ of $A^{(k-1)}$ is chosen according to some criteria. Then the ith and kth rows and the jth and kth columns of $A^{(k-1)}$ are interchanged so that $a_{ij}^{(k-1)}$ becomes $a_{kk}^{(k)}$. The ith and kth rows of $B^{(k-1)}$ are also interchanged. The pivot $a_{kk}^{(k)}$ is then used to obtain zeros in all positions in its column below the diagonal. The transformed elements are given by

$$\begin{cases} a_{ij}^{(k)} = a_{ij}^{(k-1)} & \text{for} \quad i = 1, \ldots, k \quad \text{and} \quad j = i, \ldots, n \\ b_{ij}^{(k)} = b_{ij}^{(k-1)} & \text{for} \quad i = 1, \ldots, k \quad \text{and} \quad j = 1, \ldots, r \end{cases}$$

and

$$\begin{cases} a_{ij}^{(k)} = a_{ij}^{(k-1)} - \left(\dfrac{a_{ik}^{(k-1)}}{a_{kk}^{(k)}}\right) a_{kj}^{(k)} & \text{for} \quad i = k+1, \ldots, n \\ & \text{and} \quad j = k, \ldots, n \\ b_{ij}^{(k)} = b_{ij}^{(k-1)} - \left(\dfrac{a_{ik}^{(k-1)}}{a_{kk}^{(k)}}\right) b_{kj}^{(k)} & \text{for} \quad i = k+1, \ldots, n \\ & \text{and} \quad j = 1, \ldots, r. \end{cases}$$

Also

$$a_{ij}^{(k)} = 0 \quad \text{for} \quad j = 1, \ldots, k-1 \quad \text{and} \quad i = j+1, \ldots, n.$$

Assuming that stage $k = 0$ represents the original A and B matrices, the desired triangular form is obtained at the conclusion of stage $k = n - 1$. The resulting upper triangular system is solved as shown in Section 2.3.2 preceding. This back substitution gives

$$x_{ij} = \frac{1}{a_{ii}^{(n-1)}}\left(b_{ij}^{(n-1)} - \sum_{k=i+1}^{n} a_{ik}^{(n-1)} x_{kj}\right)$$

for $j = 1, \ldots, r$ and $i = n, n-1, \ldots, 1$. Interchange of columns j and k of $A^{(k-1)}$ which was performed in order to place the pivot at the kth stage in the (k, k) diagonal position has the effect of interchanging the variables x_j and x_k in each column of solutions corresponding to a column of B. Hence before printout the jth and kth *rows* of X should be interchanged and similarly for any other column exchanges which were made on A. Row exchanges made on A were also made on B so that no further corrections are necessary.

2.3.3.2 *Positioning for Size*

Three methods of selecting the pivot $a_{ij} \neq 0$ to be used at each stage are currently in use. Other criteria are, of course, possible.

No positioning for size. Choose the pivot at the kth stage to be $a_{kk}^{(k-1)}$, the kth diagonal element of A after $(k - 1)$ transformations have been performed. Because no testing for size is done and no rows or columns are interchanged, this is the simplest and quickest method. However, no attention is paid to the relative sizes of the elements so that roundoff errors are *not* minimized. In case some $a_{kk}^{(k-1)}$ is zero (provided A is nonsingular) some nonzero element in the same column may be found that may be taken as pivot, and a row interchange is now needed.

Partial positioning for size. Choose the pivot at the kth stage to be the element of largest absolute value in column k, rows k through n. Now row interchanges are needed to place the pivot in the kth row. The tests for size and the row interchanges require more computer time, but some improvement in roundoff error should be observed.

Complete positioning for size. At the kth stage, choose the pivot as the element of largest absolute value in the entire submatrix composed of columns k through n in rows k through n. Since no two pivotal elements may lie in the same row or column, the number of elements eligible as pivots is reduced after each stage. Both row and column interchanges may now be needed in order to bring the pivot to the kth diagonal position. These interchanges together with the testing for size are time consuming, but roundoff error should be reduced considerably. Wilkinson's [108] work indicates that the error improvement may well be worth the additional work.

G. Blanch [63] has suggested that the choice of the largest element as pivot may not be superior to any other choice. Elements in the upper triangular matrix have smaller absolute error if the largest pivot is chosen each time, but the relative error of, say, the final pivot may be high, particularly in an ill-conditioned case. After back substitution, results have the relative accuracy, rather than the absolute accuracy of this pivot. Blanch suggests having all pivots the same size and equal to the nth root of the absolute value of the determinant of A in order to obtain possibly better relative errors.

Objections raised by Wilkinson [42] to choosing each pivot as the element nearest in absolute value to the nth root of the absolute value of the determinant of the matrix are the following:

1. The value of the determinant is not usually known until the Gaussian elimination has been completed.
2. There may be no elements near the prescribed value.
3. It is easy to invent examples for which it is a very poor strategy.

Wilkinson does not claim that selection of the largest elements as pivots is the best strategy, but only that no alternative practical procedure has been proposed and that complete positioning is never a very poor strategy.

The object of pivoting is to prevent loss of accuracy due to rapid increase of elements of successive $A^{(k)}$. Loss of accuracy from rapid decrease in the size of elements of successive $A^{(k)}$ is due to ill conditioning and is unavoidable.

Factors that might be considered in deciding whether to use partial or complete positioning are listed below, where $\alpha = \max |a_{ij}^{(0)}|$.

1. With partial positioning we can guarantee only that

$$|a_{ij}^{(k)}| \leq 2^{k-1}(\alpha),$$

and there exist matrices (though highly artificial) for which this bound is achieved. However, it is rare for $|a_{ij}^{(k)}|$ to be greater than 8α.

With complete positioning we have the bound

$$|a_{kk}^{(k)}| \leq \sqrt{k}(2^1 3^{1/2} 4^{1/3} \cdots k^{1/(k-1)})^{1/2}(\alpha) = f(k)(\alpha),$$

where $f(k)$ is given by the following:

k	10	20	50	100
$f(k)$	19	67	530	3300

Now $f(k) << 2^k$ and no matrix has been found for which $|a_{kk}^{(k)}| > k\alpha$. The bound is a severe overestimate.

2. On a computer with two-level store it may be easier to organize partial positioning than complete positioning.

3. For matrices having many zeros arranged in a special pattern, this pattern may be preserved by partial but destroyed by complete positioning.

4. The number of comparisons needed to select pivots for partial positioning is $n + (n-1) + (n-2) + \cdots + 1 = n(n+1)/2$, and for complete positioning is $n^2 + (n-1)^2 + (n-2)^2 + \cdots + 1^2 = \frac{1}{6}n(n+1)(2n+1)$. The extra bookkeeping required to make final permutations of rows and columns involves only two or three extra memory cells and the time to fill them.

2.3.3.3 Permutations Required

It was mentioned earlier that row interchanges to A should be accompanied by the same row interchanges to B, and column interchanges in A require row interchanges in the resulting solution matrix X (or transformed B). The reasons for these statements are easy to see. Denote by $P_{\text{row}}^{(k)}$ a matrix which performs an elementary row operation on A when multiplying A on the left at the kth stage. Denote by $I_{\text{row}}^{(k)}$ the elementary matrix that as a left multiplier interchanges two rows of A. Denote by $I_{\text{col}}^{(k)}$ the elementary matrix that interchanges two columns of A when used as a right multiplier.

Then in the case of partial positioning for size where only row operations are used, write

$$P_{\text{row}}^{(n-1)}I_{\text{row}}^{(n-1)} \cdots P_{\text{row}}^{(2)}I_{\text{row}}^{(2)}P_{\text{row}}^{(1)}I_{\text{row}}^{(1)}A = U,$$

$$Q_{\text{row}}(P_{\text{row}}^{(n-1)}I_{\text{row}}^{(n-1)} \cdots P_{\text{row}}^{(2)}I_{\text{row}}^{(2)}P_{\text{row}}^{(1)}I_{\text{row}}^{(1)}A) = I,$$

where $Q_{\text{row}} = U^{-1}$, so

$$A^{-1} = Q_{\text{row}}(P_{\text{row}}^{(n-1)} \cdots I_{\text{row}}^{(1)}).$$

The solution of $AX = B$ is then $X = A^{-1}B = Q_{\text{row}}\,(P_{\text{row}}^{(n-1)} \cdots I_{\text{row}}^{(1)})B$. Thus the row operations applied to B to produce the solution X are exactly the same row operations in the same order as used to reduce A to U and then to I.

Consider now the case of complete positioning for size where both row and column operations may be used:

$$P_{\text{row}}^{(n-1)}I_{\text{row}}^{(n-1)} \cdots I_{\text{row}}^{(2)}P_{\text{row}}^{(1)}I_{\text{row}}^{(1)}AI_{\text{col}}^{(1)}I_{\text{col}}^{(2)} \cdots I_{\text{col}}^{(n-1)} = U$$

$$Q_{\text{row}}(P_{\text{row}}^{(n-1)}I_{\text{row}}^{(n-1)} \cdots I_{\text{row}}^{(2)}P_{\text{row}}^{(1)}I_{\text{row}}^{(1)}AI_{\text{col}}^{(1)}I_{\text{col}}^{(2)} \cdots I_{\text{col}}^{(n-1)}) = I,$$

where $Q_{\text{row}} = U^{-1}$. Therefore,

$$A^{-1} = (I_{\text{col}}^{(1)} \cdots I_{\text{col}}^{(n-1)})(Q_{\text{row}})(P_{\text{row}}^{(n-1)}I_{\text{row}}^{(n-1)} \cdots P_{\text{row}}^{(1)}I_{\text{row}}^{(1)})$$

and

$$A^{-1}B = (I_{\text{col}}^{(1)} \cdots I_{\text{col}}^{(n-1)})[(Q_{\text{row}})(P_{\text{row}}^{(n-1)}I_{\text{row}}^{(n-1)} \cdots P_{\text{row}}^{(1)}I_{\text{row}}^{(1)})]B.$$

Now the operations applied to B to produce the solution X are first the same row operations in the same order as used to reduce A to U and then to I, and then the *row* permutations $(I_{\text{col}}^{(1)} \cdots I_{\text{col}}^{(n-1)})$. Since the operation $I_{\text{col}}^{(k)}$ appears as a *left* multiplier, it now acts to interchange *rows* j and k, say, where as a *right* multiplier, it interchanged *columns* j and k. These row permutations are applied in reverse order, that is, $I_{\text{col}}^{(n-1)}$ is performed first and $I_{\text{col}}^{(1)}$ is applied last.

2.3.4. Compact Methods

2.3.4.1 Crout (Banachiewicz, General Cholesky) [8, 11, 13, 18, 19, 27, 63]

The Crout method depends explicitly on the triangular resolution $A = LU$, that is, the elements of L and U are all computed and used. It is termed a "compact" scheme since the elements in the final triangular form are obtained by accumulation, dispensing with the computation and recording of intermediate $a_{ij}^{(k)}$ coefficients and reducing roundoff errors.

The method uses a fixed order of elimination and should be applied only when A is nonsingular and has been rearranged by permuting rows and columns so that all of the leading submatrices

$$\begin{bmatrix} a_{11} & \cdots & a_{1j} \\ \cdot & \cdots & \cdot \\ \cdot & \cdots & \cdot \\ \cdot & \cdots & \cdot \\ a_{j1} & \cdots & a_{jj} \end{bmatrix}$$

are nonsingular for $j = 1, \ldots, n$. See Appendix B, Theorem 63.

Setting $A = LU$, the equations for the elements of L and U from Section 2.3.1 are

$$\sum_{k=1}^{\min(i,j)} l_{ik}u_{kj} = a_{ij}.$$

Letting $u_{kk} = 1$ for $k = 1, \ldots, n$, we then have n^2 equations in n^2 unknowns, which provide the following decomposition: For $k = 1, \ldots, n$,

$$u_{kk} = 1,$$

$$l_{ik} = a_{ik} - \sum_{m=1}^{k-1} l_{im}u_{mk} \quad \text{for} \quad i = k, \ldots, n,$$

$$u_{kj} = \frac{1}{l_{kk}}\left(a_{kj} - \sum_{m=1}^{k-1} l_{km}u_{mj}\right) \quad \text{for} \quad j = k + 1, \ldots, n,$$

$$l_{ik} = 0 \quad \text{for} \quad i < k,$$

$$u_{kj} = 0 \quad \text{for} \quad j < k.$$

Hence the order of calculation is first column of L, first row of U, second column of L, second row of U, etc Note that $l_{kk} = 0$ if

$$\begin{vmatrix} a_{11} & \cdots & a_{1k} \\ \cdot & \cdots & \cdot \\ \cdot & \cdots & \cdot \\ \cdot & \cdots & \cdot \\ a_{k1} & \cdots & a_{kk} \end{vmatrix} = 0.$$

As the elements l_{ik} and u_{kj} are computed, they may be written over the A matrix in the obvious way.

After obtaining the elements of L and U, solve $AX = B$ by writing $LUX = B$, which is then equivalent to solving the triangular systems $L\xi = B$ for ξ

and then $UX = \xi$ for X. Thus

$$\xi_{ij} = \frac{1}{l_{ii}}\left(b_{ij} - \sum_{k=1}^{i-1} l_{ik}\xi_{kj}\right)$$

for $j = 1, \ldots, r$ and $i = 1, \ldots, n$,

$$x_{ij} = \left(\xi_{ij} - \sum_{k=i+1}^{n} u_{ik}x_{kj}\right)$$

for $j = 1, \ldots, r$ and $i = n, n-1, \ldots, 1$. To find A^{-1} write $A = LU$ so that $A^{-1} = U^{-1}L^{-1}$, where U^{-1} and L^{-1} are easily found as described in Section 2.3.2.

The accuracy of the method may be improved if partial pivoting is introduced. After the column l_{ik}, $i = k, \ldots, n$ is computed, the largest l_{ik} in absolute value, say l_{jk}, may be selected as l_{kk}, and its row (the jth) interchanged with the kth row in both L and A. This should cause no problems even when L and U are written over A. Then the next row of U, u_{kj} for $j = k + 1, \ldots, n$ is computed. Whenever a new column of L is computed, the largest of its elements in absolute value is chosen as diagonal.

The L and U matrices so obtained are not the triangular decomposition of A but are the decomposition of $\tilde{A} = (I_{i_n,n} \cdots I_{i_2,2}I_{i_1,1})\, A$, obtained from A by a sequence of row interchanges, $I_{i_k,k}$, of the i_k row with the kth row, where $k = 1, \ldots, n$. Hence $\tilde{A} = LU$. Now if we proceed to find $(\tilde{A})^{-1} = U^{-1}L^{-1}$, the inverse of A may then be found from

$$\tilde{A}^{-1} = A^{-1}(I_{i_1,1}^{-1}I_{i_2,2}^{-1} \cdots I_{i_n,n}^{-1}),$$

or

$$A^{-1} = \tilde{A}^{-1}(I_{i_n,n} \cdots I_{i_1,1}).$$

The matrices $I_{i_k,k}$ now act as column interchanges on $\tilde{A}^{-1}$ since they have become right multipliers. Note the order in which the column permutations are performed on $\tilde{A}^{-1}$—first $I_{i_n,n}$, then $I_{i_{n-1},n-1}$, etc. and finally $I_{i_1,1}$.

Now to solve the system $AX = B$ by the Crout method with partial pivoting, define $\tilde{A}$ as above and $\tilde{B} = (I_{i_n,n} \cdots I_{i_1,1})B$. Hence whatever row interchanges are made to L and A should also be made to B. Then we have

$$\tilde{A}\tilde{X} = \tilde{B}$$

$$\tilde{X} = \tilde{A}^{-1}\tilde{B} = A^{-1}(I_{i_1,1}^{-1} \cdots I_{i_n,n}^{-1})(I_{i_n,n} \cdots I_{i_1,1})B$$

and $\tilde{X} = A^{-1}\, B = X$. Thus the system $\tilde{A}\tilde{X} = \tilde{B}$ may be solved for $\tilde{X}$ and this is the solution, X, of $AX = B$. No column interchanges are needed.

2.3.4.2 *Symmetric Cholesky (Square-Root Method, Banachiewicz)* [7, 11, 13, 63]

Assume now that A is symmetric, that is, $A' = A$. Then by Theorem 2, Appendix B, we may write $A = LL'$ or $A = U'U$, provided A is nonsingular and has been arranged so that none of its upper left principal submatrices is singular. L is a unique lower triangular matrix, U a unique upper triangular one.

Taking $a_{ij} = a_{ji}$ and $l_{ij} = u_{ji}$ in the equations

$$\sum_{k=1}^{\min(i,j)} l_{ik}u_{kj} = a_{ij}$$

of Section 2.3.1 gives

$$\sum_{k=1}^{\min(i,j)} u_{ki}u_{kj} = a_{ij}.$$

The elements of U are given by

$$u_{11} = \sqrt{a_{11}}$$

$$u_{1j} = a_{1j}/u_{11} \qquad \text{for} \quad j = 2, \ldots, n$$

$$u_{ii} = \left(a_{ii} - \sum_{k=1}^{i-1} u_{ki}^{\,2}\right)^{1/2} \qquad \text{for} \quad i = 2, \ldots, n$$

$$u_{ij} = \frac{1}{u_{ii}}\left(a_{ij} - \sum_{k=1}^{i-1} u_{ki}u_{kj}\right) \quad \text{for} \quad j = i+1, \ldots, n \quad \text{and} \quad i = 2, \ldots, n.$$

Then $AX = U'UX = B$ and $X = (U^{-1})(U^{-1})'B$ where U^{-1} is easily found from Section 2.3.2.

The Cholesky process is particularly useful in problems of regression analysis (see [91]), in which the matrix to be decomposed is symmetric and positive definite. When A is positive definite, the u_{ii} are always real numbers since the expressions under the square-root signs are positive. To see this fact, substitute for u_{ki} in terms of the a_{ij}'s and note that the expression resulting under the square-root sign is a quotient of determinants of upper left principal minors of A which are positive if A is positive definite. Also see Appendix B, Theorems 20, 21, and 22.

The decomposition may still be obtained for symmetric A even if A is not positive definite, but the diagonal elements of U may be complex numbers.

No pivoting is required in the case where A is symmetric and positive definite. If A is symmetric but not positive definite, pivoting should be used. See Chapters 4 and 5.

2.3.4.3 *Cholesky for Hermitian Matrices* [27, 84]

For the case where A is Hermitian, write $A = L(L^*)'$ or $A = (U^*)'U$ (see Theorem 3, Appendix B), assuming again that A is nonsingular and arranged so that none of its upper left principal submatrices is singular.

Taking $a_{ij} = \bar{a}_{ji}$ and $l_{ij} = \bar{u}_{ji}$ in the equations of Section 2.3.1, we have

$$\sum_{k=1}^{\min(i,j)} \bar{u}_{ki}u_{kj} = a_{ij},$$

from which the following equations are derived:

$$\bar{u}_{11}u_{11} = a_{11}$$

$$\bar{u}_{ii}u_{ii} = \left(a_{ii} - \sum_{k=1}^{i-1} \bar{u}_{ki}u_{ki}\right) \quad \text{for} \quad i = 1, \ldots, n$$

$$u_{ij} = \frac{1}{\bar{u}_{ii}}\left(a_{ij} - \sum_{k=1}^{i-1} \bar{u}_{ki}u_{kj}\right) \quad \text{for} \quad i = 1, \ldots, n \quad \text{and} \quad j = i + 1, \ldots, n.$$

Note that since A is Hermitian, $a_{ii} = \bar{a}_{ii}$, so that a_{ii} is real. Also $\bar{u}_{ii}u_{ii} = |u_{ii}|^2$ is real and positive, as is $\bar{u}_{ki}u_{ki} = |u_{ki}|^2$. If A is positive definite, the right-hand sides of the first two formulas above are positive so that u_{ii} may be taken real and positive, that is,

$$u_{ii} = \left(a_{ii} - \sum_{k=1}^{i-1} |u_{ki}|^2\right)^{1/2} \quad \text{for} \quad i = 1, \ldots, n$$

$$u_{ij} = \frac{1}{u_{ii}}\left(a_{ij} - \sum_{k=1}^{i-1} \bar{u}_{ki}u_{kj}\right) \quad \text{for} \quad i = 1, \ldots, n \quad \text{and} \quad j = i + 1, \ldots, n.$$

2.3.4.4 *Doolittle* (*Black*) [7, 13, 27, 63]

The Doolittle method is another compact elimination scheme where the pivots are taken in succession along the main diagonal. Elimination is done by rows until a final triangular system is produced. Elements in the final triangular form are obtained by accumulation thus reducing roundoff errors. Back substitution is then used. The only recorded quantities are the elements of L and U and the results of the back substitution. Intermediate coefficients $a_{ij}^{(k)}$ are not computed and recorded as in Gaussian elimination. If the rows of the final upper triangular form U are divided by the diagonal elements, the upper triangular form of Crout is obtained. It is again assumed A is nonsingular and so arranged that all leading upper left submatrices are nonsingular.

Set $A = LU$ and assume L to have unit diagonal to obtain

$$l_{11} = 1$$

$$u_{1j} = a_{1j} \quad \text{for} \quad j = 1, \ldots, n,$$

and for $k = 2, \ldots, n$,

$$l_{kk} = 1$$

$$l_{kj} = \frac{1}{u_{jj}}\left(a_{kj} - \sum_{m=1}^{j-1} l_{km}u_{mj}\right) \quad \text{for} \quad j = 1, \ldots, k-1,$$

$$u_{kj} = \left(a_{kj} - \sum_{m=1}^{k-1} l_{km}u_{mj}\right) \qquad \text{for} \quad j = k, \ldots, n.$$

The order of calculation is first row of L, first row of U, second row of L, second row of U, etc. Those elements of L and U not computed are, of course, zero. Once L and U have been found, $L\xi = B$ is solved for ξ and $UX = \xi$ for X, exactly as given in Sections 2.3.4.1 and 2.3.2.

Pivoting should be used to improve accuracy (see Chapter 4).

2.3.5 Below-the-Line Methods [13, 60, 63]

2.3.5.1 Crout's Below-the-Line Method

Write

$$M = \left(\begin{array}{c|c} A & K \\ \hline H & 0 \end{array}\right)$$

where A is $n \times n$, K is $n \times p$, and H is $q \times n$.

Then the result of applying the Crout algorithm to the first n rows and columns of M (which decomposes A to $A = LDU$) is the matrix

$$\left(\begin{array}{c|c} \begin{matrix} & \tilde{U} \\ LD & \end{matrix} & (LD)^{-1}K \\ \hline HU^{-1} & \text{blank} \end{array}\right),$$

where the dimensions of the four blocks are the same as before, and where $\tilde{U}$ is U with zero diagonal. If we let $H = K = I$, then U^{-1} and $(LD)^{-1}$ are obtained; or, if $K = B$ and $H = I$, then U^{-1} and $(LD)^{-1}B$ for the solution $X = U^{-1}(LD)^{-1}B$ of $AX = LDUX = B$ are obtained.

2.3.5.2 Aitken's Below-the-Line Method

Again write

$$M = \left(\begin{array}{c|c} A & K \\ \hline H & 0 \end{array}\right)$$

where A, K, and H have the same dimensions as above. Choosing pivots from A and using Gaussian elimination on the rows of A and H, A becomes upper triangular U, and H becomes zero. Then the lower right corner which was zero now becomes $-HA^{-1}K$, and K becomes some matrix $F = UA^{-1}K$. In particular let $K = I$ and $H = -I$, then the lower right corner becomes A^{-1}. If we let $K = B$ and $H = -I$, then the lower right corner becomes $A^{-1}B$.

2.4 METHODS BASED ON DIAGONALIZATION

2.4.1 Gauss–Jordan Elimination [8, 11, 13, 19, 22, 59, 63]

Any matrix A is equivalent to a diagonal matrix (Theorem 8, Appendix B) and this diagonal matrix is obtained from A by a finite number of elementary operations on the rows and/or columns of A. If A is real, so is the resulting diagonal matrix. The method is similar to Gaussian elimination, but now the elements above the diagonal are eliminated as well, so that back substitution is no longer necessary.

Pivots are chosen by one of the methods suggested in the Gaussian elimination, Section 2.3.3.2, and, assuming that at the kth stage the chosen pivot is moved to the kth diagonal position by appropriate row and column interchanges, the result of the elimination is

$$a_{ij}^{(k)} = a_{ij}^{(k-1)} - \left(\frac{a_{ik}^{(k-1)}}{a_{kk}^{(k-1)}}\right) a_{kj}^{(k-1)} \quad \text{for} \quad i = 1, \ldots, k-1, k+1, \ldots, n \quad \text{and} \quad j = k, \ldots, n$$

$$b_{ij}^{(k)} = b_{ij}^{(k-1)} - \left(\frac{a_{ik}^{(k-1)}}{a_{kk}^{(k-1)}}\right) b_{kj}^{(k-1)} \quad \text{for} \quad i = 1, \ldots, k-1, k+1, \ldots, n \quad \text{and} \quad j = 1, \ldots, r$$

$$a_{ij}^{(k)} = a_{ij}^{(k-1)} \quad \text{for} \quad i = 1, \ldots, n \quad \text{and} \quad j = 1, \ldots, k-1.$$

If stage $k = 0$ represents the original A and B matrices, the desired diagonal form is obtained at the conclusion of stage $k = n - 1$. The solutions of $AX = B$ are then

$$x_{ij} = \frac{b_{ij}^{(n-1)}}{a_{ii}^{(n-1)}} \quad \text{for} \quad i = 1, \ldots, n \quad \text{and} \quad j = 1, \ldots, r.$$

The LDU representation is bypassed in this process since the order of operations reduces an entire column at a time and no intermediate upper triangular form is obtained. As in Gaussian elimination, all row operations performed on A are at the same time performed on B, and any column interchanges of A require corresponding row interchanges in the final $A^{-1}B$.

Note that it is not actually necessary to move the pivot to the kth diagonal position before performing the eliminations. All that is required is that some record be kept of those rows and columns which have been used. The final transformed A should then have one nonzero element in each row and column, so that a sequence of row transformations will always put it into diagonal form. The same row transformation should also be applied to B.

Assume that in solving $AX = B$, storage is provided for the $n \times n$ matrix A and the $n \times r$ matrix B, where $r \geq n$. Then, at the conclusion of the Gauss–Jordan process, the A storage contains a diagonal matrix which is easily reduced to the identity matrix I, and the B storage contains $X = A^{-1}B$. If $B = I$, then the B storage contains $X = A^{-1}$ at the conclusion of the process. Thus either A^{-1} or $A^{-1}B$ is obtained, but not both unless the B storage is appended so as to include B followed by I. If desired, both A^{-1} and $A^{-1}B$ could be obtained if we allow the A matrix to be overwritten one column at a time {or alternatively, append one column to the A matrix (see [34]) and operate on it concurrently with A}. The column j to be overwritten is, of course, the one currently being annihilated by pivot a_{ij}. It would normally be left with only one nonzero element. Instead, this column is replaced by column i of the transformed identity matrix. Division of row i of A, B, or I by its pivot a_{ij} can be performed as the pivots are selected or pivots can be saved and the division done at the conclusion of all the eliminations. Finally, row interchanges on A and B are performed corresponding to those interchanges necessary to put A in diagonal form; column interchanges on A are necessary to produce A^{-1}. The latter are required because column i (rather than column j) of the transformed identity matrix was written into column j of A. If the pivots selected are $a_{i_1,j_1}, a_{i_2,j_2}, \ldots, a_{i_n,j_n}$, then the following row changes must be made on the transformed A and B:

old row i_1 becomes new row j_1
old row i_2 becomes new row j_2
$\cdot \qquad \cdot$
$\cdot \qquad \cdot$
$\cdot \qquad \cdot$
old row i_n becomes new row j_n.

Also the following column interchanges must be made on the final overwritten A:

old column j_1 becomes new column i_1
old column j_2 becomes new column i_2
$\cdot \qquad \cdot$
$\cdot \qquad \cdot$
$\cdot \qquad \cdot$
old column j_n becomes new column i_n.

TABLE 2.1
Transformations

A	I	Overwritten A	Pivot
$\begin{pmatrix} +2 & 4 & -2 \\ 2 & 2 & 3 \\ -1 & -1 & +1 \end{pmatrix}$	$\begin{pmatrix} 1 & 0 & 0 \\ 0 & 1 & 0 \\ 0 & 0 & 1 \end{pmatrix}$	—	$a_{12}^{(0)} = 4$
$\begin{pmatrix} +\frac{1}{2} & 1 & -\frac{1}{2} \\ 2 & 2 & 3 \\ -1 & -1 & +1 \end{pmatrix}$	$\begin{pmatrix} \frac{1}{4} & 0 & 0 \\ 0 & 1 & 0 \\ 0 & 0 & 1 \end{pmatrix}$	—	
$\begin{pmatrix} +\frac{1}{2} & 1 & -\frac{1}{2} \\ 1 & 0 & 4 \\ -\frac{1}{2} & 0 & +\frac{1}{2} \end{pmatrix}$	$\begin{pmatrix} \frac{1}{4} & 0 & 0 \\ -\frac{1}{2} & 1 & 0 \\ \frac{1}{4} & 0 & 1 \end{pmatrix}$	$\begin{pmatrix} +\frac{1}{2} & \frac{1}{4} & -\frac{1}{2} \\ 1 & -\frac{1}{2} & 4 \\ -\frac{1}{2} & \frac{1}{4} & +\frac{1}{2} \end{pmatrix}$	$a_{23}^{(1)} = 4$
$\begin{pmatrix} +\frac{1}{2} & 1 & -\frac{1}{2} \\ \frac{1}{4} & 0 & 1 \\ -\frac{1}{2} & 0 & +\frac{1}{2} \end{pmatrix}$	$\begin{pmatrix} \frac{1}{4} & 0 & 0 \\ -\frac{1}{8} & \frac{1}{4} & 0 \\ \frac{1}{4} & 0 & 1 \end{pmatrix}$	$\begin{pmatrix} \frac{1}{2} & \frac{1}{4} & -\frac{1}{2} \\ \frac{1}{4} & -\frac{1}{8} & 1 \\ -\frac{1}{2} & \frac{1}{4} & \frac{1}{2} \end{pmatrix}$	
$\begin{pmatrix} \frac{5}{8} & 1 & 0 \\ \frac{1}{4} & 0 & 1 \\ -\frac{5}{8} & 0 & 0 \end{pmatrix}$	$\begin{pmatrix} \frac{3}{16} & \frac{1}{8} & 0 \\ -\frac{1}{8} & \frac{1}{4} & 0 \\ \frac{5}{16} & -\frac{1}{8} & 1 \end{pmatrix}$	$\begin{pmatrix} \frac{5}{8} & \frac{3}{16} & \frac{1}{8} \\ \frac{1}{4} & -\frac{1}{8} & \frac{1}{4} \\ -\frac{5}{8} & \frac{5}{16} & -\frac{1}{8} \end{pmatrix}$	$a_{31}^{(2)} = -\frac{5}{8}$
$\begin{pmatrix} \frac{5}{8} & 1 & 0 \\ \frac{1}{4} & 0 & 1 \\ 1 & 0 & 0 \end{pmatrix}$	$\begin{pmatrix} \frac{3}{16} & \frac{1}{8} & 0 \\ -\frac{1}{8} & \frac{1}{4} & 0 \\ -\frac{1}{2} & +\frac{1}{5} & -\frac{8}{5} \end{pmatrix}$	$\begin{pmatrix} \frac{5}{8} & \frac{3}{16} & \frac{1}{8} \\ \frac{1}{4} & -\frac{1}{8} & \frac{1}{4} \\ 1 & -\frac{1}{2} & \frac{1}{5} \end{pmatrix}$	
$\begin{pmatrix} 0 & 1 & 0 \\ 0 & 0 & 1 \\ 1 & 0 & 0 \end{pmatrix}$	$\begin{pmatrix} \frac{1}{2} & 0 & 1 \\ 0 & \frac{1}{5} & \frac{2}{5} \\ -\frac{1}{2} & \frac{1}{5} & -\frac{8}{5} \end{pmatrix}$	$\begin{pmatrix} 1 & \frac{1}{2} & 0 \\ \frac{2}{5} & 0 & \frac{1}{5} \\ -\frac{8}{5} & -\frac{1}{2} & \frac{1}{5} \end{pmatrix}$	
$\begin{pmatrix} 1 & 0 & 0 \\ 0 & 1 & 0 \\ 0 & 0 & 1 \end{pmatrix}$	$\begin{pmatrix} -\frac{1}{2} & \frac{1}{5} & -\frac{8}{5} \\ \frac{1}{2} & 0 & 1 \\ 0 & \frac{1}{5} & \frac{2}{5} \end{pmatrix}$	$\begin{pmatrix} -\frac{8}{5} & -\frac{1}{2} & \frac{1}{5} \\ 1 & \frac{1}{2} & 0 \\ \frac{2}{5} & 0 & \frac{1}{5} \end{pmatrix}$	Row Permutations old new 1 → 2 2 → 3 3 → 1
	$A^{-1} = \begin{pmatrix} -\frac{1}{2} & \frac{1}{5} & -\frac{8}{5} \\ \frac{1}{2} & 0 & 1 \\ 0 & \frac{1}{5} & \frac{2}{5} \end{pmatrix}$		Column Permutations on Overwritten A old new 2 → 1 3 → 2 1 → 3.

A simple example is in order. Three matrices are given simultaneously: the A matrix, the identity matrix I, and the overwritten A matrix, so that comparisons may be made at each stage of the transformation. No B matrix is given, but operations upon B are the same as those upon A, and involve only row operations. (See Table 2.1.)

Suppose, alternatively, that the kth pivot is actually moved to the kth diagonal position before eliminations are performed. Then at stage k, if $a_{ij}^{(k)}$ is the pivot selected, we interchange rows i and k, and interchange columns j and k. Any row interchanges made on A must be made simultaneously on I and on the overwritten A. Any column interchanges made on A must be made simultaneously on the overwritten A. When the eliminations have been completed and the A matrix has been reduced to the identity, the corresponding I and overwritten A matrices must undergo row interchanges corresponding to the column interchanges performed on A, but in reverse order (see Section 2.3.3.3). The procedure is illustrated by Table 2.2.

2.4.2 Product Form of the Inverse [86]

Once the basic mathematical representation has been provided for a matrix inversion, the actual manner in which the process is programmed for a computer, as well as the characteristics of the computer, determine its efficiency, accuracy, and storage requirements.

An example is provided by a technique for the Gauss–Jordan reduction known as the "product form of the inverse." If the purpose of obtaining A^{-1} is only to solve a number of cases $Ax = b$ for different b vectors, then it is pointless to waste computing time on A^{-1} when a special array of multipliers will do. This array, the "product form of the inverse" is commonly used in linear programming routines in which many systems are to be solved which differ from each other by only one of a small number of columns.

Define the following iteration for finding A^{-1}. Set $W^{(0)} = A$ and let the matrix $C^{(k)} = (C_{ij}^{(k)})$ be defined as

$$
\begin{aligned}
C_{ii}^{(k)} &= 1 && \text{for} \quad i \neq k \\
C_{ij}^{(k)} &= 0 && \text{for} \quad j \neq k \\
C_{kk}^{(k)} &= \frac{1}{W_{kk}^{(k-1)}} \\
C_{ik}^{(k)} &= \frac{-W_{ik}^{(k-1)}}{W_{kk}^{(k-1)}} && \text{for} \quad i \neq k.
\end{aligned}
$$

$C^{(k)}$ is the identity matrix except for the kth column. Then

$$W^{(k)} = C^{(k)} W^{(k-1)} = C^{(k)} C^{(k-1)} \cdots C^{(1)} A$$

TABLE 2.2
Transformations

A	I	Overwritten A	Pivot
$\begin{pmatrix} 2 & 4 & -2 \\ 2 & 2 & 3 \\ -1 & -1 & 1 \end{pmatrix}$	$\begin{pmatrix} 1 & 0 & 0 \\ 0 & 1 & 0 \\ 0 & 0 & 1 \end{pmatrix}$		$a_{12}^{(0)} = 4$
$\begin{pmatrix} 4 & 2 & -2 \\ 2 & 2 & 3 \\ -1 & -1 & 1 \end{pmatrix}$	$\begin{pmatrix} 1 & 0 & 0 \\ 0 & 1 & 0 \\ 0 & 0 & 1 \end{pmatrix}$		Columns 1 and 2 interchanged in A.
$\begin{pmatrix} 1 & \frac{1}{2} & -\frac{1}{2} \\ 2 & 2 & 3 \\ -1 & -1 & 1 \end{pmatrix}$	$\begin{pmatrix} \frac{1}{4} & 0 & 0 \\ 0 & 1 & 0 \\ 0 & 0 & 1 \end{pmatrix}$		Divide row 1 by $a_{11}^{(0)} = 4$
$\begin{pmatrix} 1 & \frac{1}{2} & -\frac{1}{2} \\ 0 & 1 & 4 \\ 0 & -\frac{1}{2} & \frac{1}{2} \end{pmatrix}$	$\begin{pmatrix} \frac{1}{4} & 0 & 0 \\ -\frac{1}{2} & 1 & 0 \\ \frac{1}{4} & 0 & 1 \end{pmatrix}$	$\begin{pmatrix} \frac{1}{4} & \frac{1}{2} & -\frac{1}{2} \\ -\frac{1}{2} & 1 & 4 \\ \frac{1}{4} & -\frac{1}{2} & \frac{1}{2} \end{pmatrix}$	$a_{23}^{(1)} = 4$
$\begin{pmatrix} 1 & -\frac{1}{2} & \frac{1}{2} \\ 0 & 4 & 1 \\ 0 & \frac{1}{2} & -\frac{1}{2} \end{pmatrix}$	$\begin{pmatrix} \frac{1}{4} & 0 & 0 \\ -\frac{1}{2} & 1 & 0 \\ \frac{1}{4} & 0 & 1 \end{pmatrix}$	$\begin{pmatrix} \frac{1}{4} & -\frac{1}{2} & \frac{1}{2} \\ -\frac{1}{2} & 4 & 1 \\ \frac{1}{4} & \frac{1}{2} & -\frac{1}{2} \end{pmatrix}$	Interchange columns 2 and 3 in A and overwritten A.
$\begin{pmatrix} 1 & -\frac{1}{2} & \frac{1}{2} \\ 0 & 1 & \frac{1}{4} \\ 0 & \frac{1}{2} & -\frac{1}{2} \end{pmatrix}$	$\begin{pmatrix} \frac{1}{4} & 0 & 0 \\ -\frac{1}{8} & \frac{1}{4} & 0 \\ \frac{1}{4} & 0 & 1 \end{pmatrix}$	$\begin{pmatrix} \frac{1}{4} & -\frac{1}{2} & \frac{1}{2} \\ -\frac{1}{8} & 1 & \frac{1}{4} \\ \frac{1}{4} & \frac{1}{2} & -\frac{1}{2} \end{pmatrix}$	Divide row 2 by $a_{22}^{(1)} = 4$.
$\begin{pmatrix} 1 & 0 & \frac{5}{8} \\ 0 & 1 & \frac{1}{4} \\ 0 & 0 & -\frac{5}{8} \end{pmatrix}$	$\begin{pmatrix} \frac{3}{16} & \frac{1}{8} & 0 \\ -\frac{1}{8} & \frac{1}{4} & 0 \\ \frac{5}{16} & -\frac{1}{8} & 1 \end{pmatrix}$	$\begin{pmatrix} \frac{3}{16} & \frac{1}{8} & \frac{5}{8} \\ -\frac{1}{8} & \frac{1}{4} & \frac{1}{4} \\ \frac{5}{16} & -\frac{1}{8} & -\frac{5}{8} \end{pmatrix}$	$a_{33}^{(2)} = -\frac{5}{8}$
$\begin{pmatrix} 1 & 0 & \frac{5}{8} \\ 0 & 1 & \frac{1}{4} \\ 0 & 0 & 1 \end{pmatrix}$	$\begin{pmatrix} \frac{3}{16} & \frac{1}{8} & 0 \\ -\frac{1}{8} & \frac{1}{4} & 0 \\ -\frac{1}{2} & \frac{1}{5} & -\frac{8}{5} \end{pmatrix}$	$\begin{pmatrix} \frac{3}{16} & \frac{1}{8} & \frac{5}{8} \\ -\frac{1}{8} & \frac{1}{4} & \frac{1}{4} \\ -\frac{1}{2} & \frac{1}{5} & 1 \end{pmatrix}$	Divide row 3 by $a_{33}^{(2)} = -\frac{5}{8}$.
$\begin{pmatrix} 1 & 0 & 0 \\ 0 & 1 & 0 \\ 0 & 0 & 1 \end{pmatrix}$	$\begin{pmatrix} \frac{1}{2} & 0 & 1 \\ 0 & \frac{1}{5} & \frac{2}{5} \\ -\frac{1}{2} & \frac{1}{5} & -\frac{8}{5} \end{pmatrix}$	$\begin{pmatrix} \frac{1}{2} & 0 & 1 \\ 0 & \frac{1}{5} & \frac{2}{5} \\ -\frac{1}{2} & \frac{1}{5} & -\frac{8}{5} \end{pmatrix}$	Interchange rows 2 and 3, then interchange rows 1 and 2 to obtain the inverse.
	$A^{-1} = \begin{pmatrix} -\frac{1}{2} & \frac{1}{5} & -\frac{8}{5} \\ \frac{1}{2} & 0 & 1 \\ 0 & \frac{1}{5} & \frac{2}{5} \end{pmatrix}$		Inverse.

and

$$W^{(n)} = I = C^{(n)} \cdots C^{(1)}A$$

so

$$A^{-1} = C^{(n)} \cdots C^{(1)},$$

the product form of the inverse. Since the first $(k-1)$ columns of $W^{(k-1)}$ are corresponding unit vectors, the multiplication $C^{(k)}W^{(k-1)}$ does not affect these columns and need not be performed, that is,

$$W_{ij}^{(k)} = W_{ij}^{(k-1)} + C_{ik}^{(k)}W_{kj}^{(k-1)} \quad \text{all} \quad j \geq k, \quad \text{all} \quad i \neq k$$

$$W_{kj}^{(k)} = C_{kk}^{(k)}W_{kj}^{(k-1)} \quad \text{all} \quad j \geq k.$$

It has been assumed here that the pivots are taken in succession along the main diagonal. Permutations of rows or columns necessary for other choice of pivots are described in the preceding Sections 2.3.3.2, 2.3.3.3, and 2.4.1, and in great detail in [86].

2.4.3 Method of Congruent Transformations [51]

Assume that the given matrix A is symmetric. Then by Theorem 9, Appendix B, A may be converted to diagonal form by means of a congruent (actually an orthogonal) transformation, that is, $D = Q'AQ$.

After the matrix A is transformed to diagonal form D by a sequence of congruent transformations, the diagonal matrix is then inverted and the result is submitted to the reverse sequence of congruent transformations.

Let $A^{(1)} = A$ and $A^{(i+1)} = (Q^{(i)})'A^{(i)}Q^{(i)}$ be the result of i congruent transformations performed on A to reduce the off-diagonal elements of the first i rows and columns of $A^{(i+1)}$ to zero. Define

$$Q^{(i)} = (q_{jk}^{(i)}) = \delta_{jk}$$

$$\text{except that } q_{ik}^{(i)} = -a_{ik}^{(i)}/a_{ii}^{(i)} \quad \text{for} \quad k > i.$$

Choose the ith pivot as the largest diagonal element $a_{pp}^{(i)}, p \geq i$ in the matrix $A^{(i)}$ being reduced, and interchange row i with row p and, to preserve symmetry, column i with column p. The foregoing interchanges may be represented by the congruent transformation matrix

$$E^{(i)} = (e_{jk}^{(i)}) = \begin{matrix} \delta_{jk} \quad \text{except that} \\ e_{ip}^{(i)} = e_{pi}^{(i)} = 1 \quad \text{and} \quad e_{ii}^{(i)} = e_{pp}^{(i)} = 0 \end{matrix}$$

which is used as both a right and a left multiplier. The reduction is then performed. After $(n-1)$ such interchanges and reductions, a diagonal matrix $A^{(n)}$ is obtained.

$$A^{(n)} = (Q^{(n-1)})'E^{(n-2)}(Q^{(n-2)})' \cdots (Q^{(1)})'E^{(1)}A^{(1)}E^{(1)} \times Q^{(1)} \cdots Q^{(n-2)}E^{(n-2)}Q^{(n-1)}.$$

It then follows that

$$A^{-1} = (A^{(1)})^{-1} = E^{(1)}Q^{(1)}E^{(2)} \cdots E^{(n-2)}Q^{(n-1)}(A^{(n)})^{-1} \times (Q^{(n-1)})'E^{(n-2)} \cdots E^{(2)}(Q^{(1)})'E^{(1)}.$$

If at some stage all diagonal elements $a_{jj}^{(i)}$, $j \geq i$ have vanished, then the method will fail. This difficulty cannot occur if A is positive definite, must occur if A is singular, and may or may not occur if A is negative definite.

2.5 ORTHOGONALIZATION [13, 19, 27, 60, 63, 73]

Although the number of arithmetic operations involved in these orthogonal methods is quite unfavorable (see Table 7.1) the methods are still of interest.

2.5.1 Method of Orthogonal Vectors

2.5.1.1 Symmetric Case

Let A be symmetric. We wish to solve $Ax = b$ and proceed by expressing the solution column vector x as a linear combination of n specially chosen column vectors l_r so that

$$x = \sum_{r=1}^{n} c_r l_r.$$

The vectors l_r are chosen to be A-orthogonal, that is, $l_s' A l_r = 0$ for $r \neq s$. Since A is assumed symmetric, it follows also that $l_s' A l_r = l_r' A l_s = 0$ for $r \neq s$. Then

$$Ax = \sum_{r=1}^{n} c_r A l_r \quad \text{and} \quad l_s' A x = \sum_{r=1}^{n} c_r l_s' A l_r = l_s' b,$$

or

$$c_s l_s' A l_s = l_s' b \quad \text{and} \quad c_s = l_s' b (l_s' A l_s)^{-1} \quad \text{for} \quad s = 1, \ldots, n.$$

To produce the desired vectors l_r, start with n linearly independent vectors y_i, as for example, the unit vectors. Hence take y_i as the vector with 1 in the ith component, zero elsewhere.

Set

$$\begin{aligned} l_1 &= y_1 \\ l_2 &= y_2 + \lambda_{12} l_1 \\ &\vdots \\ l_r &= y_r + \sum_{i=1}^{r-1} \lambda_{ir} l_i. \end{aligned}$$

Since it is required that

$$l_s' A l_r = l_s' A \left(y_r + \sum_{i=1}^{r-1} \lambda_{ir} l_i \right) = l_s' A y_r + l_s' A \lambda_{sr} l_s = 0$$

for $s < r$, take λ_{sr} as

$$\lambda_{sr} = -(l_s' A l_s)^{-1} (l_s' A y_r).$$

Define L as the lower unit triangular matrix whose rows are the vectors $l_1, \ldots, l_n$ and L' as the upper unit triangular matrix whose columns are the vectors $l_1, \ldots, l_n$. Then, since the vectors l_r are A-orthogonal, write

$$LAL' = D,$$

where D is a diagonal matrix. Hence

$$A = L^{-1}D(L^{-1})' = L^{-1}D(L')^{-1}.$$

Because L^{-1} is also lower triangular, the decomposition of Theorem 2, Appendix B, is again obtained. Now

$$A^{-1} = L'D^{-1}L$$

and

$$x = A^{-1}b = (L'D^{-1}L)b.$$

Also

$$x = \sum_{r=1}^{n} c_r l_r = \sum_{r=1}^{n} l_r' b(l_r'Al_r)^{-1} l_r = \left[\sum_{r=1}^{n} l_r'(l_r'Al_r)^{-1} l_r\right] b.$$

2.5.1.2 Nonsymmetric Case

For the nonsymmetric case an analogous process will give simultaneously the solutions of two sets of linear equations

$$Ax = b^{(1)} \quad \text{and} \quad A'z = b^{(2)}.$$

Two biconjugate sets of vectors $x^{(r)}$ and $z^{(r)}$ are computed from $y^{(r)}$, the unit vectors, say, using

$$\lambda_{rs} = \frac{(z^{(r)}, Ay^{(s)})}{(z^{(r)}, Ax^{(r)})} \quad \text{for the } x^{(r)}\text{'s}$$

and

$$\tilde{\lambda}_{rs} = \frac{(x^{(r)}, A'y^{(s)})}{(x^{(r)}, A'z^{(r)})} \quad \text{for the } z^{(r)}\text{'s}.$$

Then write

$$x = \sum_{r=1}^{n} \alpha_r x^{(r)} \quad \text{and} \quad z = \sum_{r=1}^{n} \beta_r z^{(r)},$$

where

$$\alpha_r = \frac{(z^{(r)}, b^{(1)})}{(z^{(r)}, Ax^{(r)})}$$

and

$$\beta_r = \frac{(x^{(r)}, b^{(2)})}{(x^{(r)}, A'z^{(r)})}.$$

In matrix notation $Z'AX = X'AZ = D$, where

$X = (x^{(r)})$	is a unit upper triangular matrix,
$Z = (z^{(r)})$	is a unit upper triangular matrix,
$C_1 = (\lambda_{rs})$	is a unit upper triangular matrix and $C_1 = X^{-1}$,
$C_2 = (\tilde{\lambda}_{rs})$	is a unit upper triangular matrix and $C_2 = Z^{-1}$,
$c^{(1)} = (\alpha_r)$	is a vector and $c^{(1)} = D^{-1}Z'b^{(1)}$,
$c^{(2)} = (\beta_r)$	is a vector and $c^{(2)} = D^{-1}X'b^{(2)}$.

Thus

$$x = Xc^{(1)} = XD^{-1}Z'b^{(1)} = X(Z'AX)^{-1}Z'b^{(1)} = A^{-1}b^{(1)}$$

$$z = Zc^{(2)} = ZD^{-1}X'b^{(2)} = Z(X'A'Z)^{-1}X'b^{(2)} = (A')^{-1}b^{(2)}.$$

2.5.2 Method of Matrix Orthogonalization

2.5.2.1 Method 1

Let $AX = B$ where A is real and nonsingular. By Schmidt's theorem, Theorem 19, Appendix B, there is a nonsingular triangular matrix, say L^{-1}, which transforms A into an orthogonal matrix, R.

Write $L^{-1}A = R$ or $A = LR$, then $LRX = B$ or $X = R^{-1}L^{-1}B$. Now, since R is orthogonal, $RR' = D$, or $R^{-1} = R'D^{-1}$, where D is diagonal. Hence $X = (R'D^{-1}L^{-1})B$.

Note that we could take $\phi = (D^{-1/2}R)$ with $\phi\phi' = D^{-1/2}RR'D^{-1/2} = I$. However, $D^{-1/2}$ need not have real elements unless A is positive definite.

The classical method for finding R is a Gram–Schmidt orthogonalization procedure which has been used previously to produce A-orthogonal vectors from a set of linearly independent ones. Let $R = (r_{ij})$ and where

$$\begin{aligned} r_{1j} &= a_{1j} \\ r_{2j} &= a_{2j} - l_{21}r_{1j} \\ &\;\;\vdots \\ r_{ij} &= a_{ij} - \sum_{k=1}^{i-1} l_{ik}r_{kj} \end{aligned}$$

and

$$l_{ik} = \sum_{j=1}^{n} \left(\frac{r_{kj}a_{ij}}{r_{kj}^{\;2}}\right) \qquad \text{for} \quad k = 1, \ldots, i-1.$$

Now define $l_{ii} = 1$ for $i = 1, \ldots, n$. Then L is unit lower triangular, $A = LU$ and $RR' = D$. Alternatively, we could find a unit upper triangular matrix, U, and an orthogonal matrix R, such that $A = RU$, $R'R = D$ and $X = (RU)^{-1}B = (U^{-1}D^{-1}R')B$.

2.5.2.2 *Method 2*

Also by Schmidt's theorem, if A is any nonsingular matrix, there exists a nonsingular triangular matrix L^{-1} that transforms A into a unitary matrix V. Thus $L^{-1}A = V$ or $A = LV$ and $V(V^*)' = D$, where D is diagonal. Then the solution of $AX = B$ is $X = A^{-1}B = V^{-1}L^{-1}B = (V^*)'D^{-1}L^{-1}B$. Let $V = (v_{ij})$. Define

$$v_{1j} = a_{1j}; \qquad v_{ij} = a_{ij} \sum_{k=1}^{i-1} l_{ik} v_{kj},$$

where l_{ik} is determined so that $V(V^*)' = D$, that is, so that

$$\sum_{k=1}^{n} v_{ik}\bar{v}_{ik} \neq 0, \quad \text{but} \quad \sum_{k=1}^{n} v_{ik}\bar{v}_{jk} = 0 \quad \text{for} \quad i \neq j.$$

Hence

$$\sum_{k=1}^{n} \left(a_{ik} - \sum_{p=1}^{i-1} l_{ip} v_{pk} \right) \bar{v}_{jk} = \sum_{k=1}^{n} a_{ik}\bar{v}_{jk} - \sum_{k=1}^{n} \sum_{p=1}^{i-1} l_{ip} v_{pk} \bar{v}_{jk} = 0.$$

Since $v_{pk}\bar{v}_{jk} = 0$ for $p \neq j$ and $p < i, j < i$, it follows that

$$\sum_{k=1}^{n} (a_{ik}\bar{v}_{jk} - l_{ij} v_{jk}\bar{v}_{jk}) = 0,$$

or

$$l_{ij} = \frac{\sum_{k=1}^{n} a_{ik}\bar{v}_{jk}}{\sum_{k=1}^{n} (v_{jk}\bar{v}_{jk})} \quad \text{for} \quad j < i.$$

2.5.2.3 *Method 3*

A further matrix orthogonalization process is based on the fact that the rows of A^{-1} are orthogonal to the columns of A. Thus successive matrices $B^{(0)}, B^{(1)}, \ldots, B^{(n)}$ are computed beginning with an arbitrary matrix $B^{(0)}$ so that the rows of $B^{(i)}$ are orthogonal to the first i columns of A.

Let $A = (a_{kj})$ and $B^{(i)} = (b_{kj}^{(i)})$ for $i, k, j = 1, \ldots, n$. Define scalars α_{ki} for $k \neq i$:

$$\alpha_{ki} = \frac{(\text{row } k \text{ of } B^{(i-1)})(\text{column } i \text{ of } A)}{(\text{row } i \text{ of } B^{(i-1)})(\text{column } i \text{ of } A)} = \frac{\sum_{j=1}^{n} b_{kj}^{(i-1)} a_{ji}}{\sum_{j=1}^{n} b_{ij}^{(i-1)} a_{ji}}.$$

Then the matrices $B^{(i)}$ are defined recursively for $i = 1, \ldots, n$:

row i of $B^{(i)}$ = row i of $B^{(i-1)}$,

row k of $B^{(i)}$ = row k of $B^{(i-1)} - \alpha_{ki}$(row i of $B^{(i)}$) for $k \neq i$.

In terms of individual matrix elements, the equivalent definition is

$$b_{ij}^{(i)} = b_{ij}^{(i-1)},$$
$$b_{kj}^{(i)} = b_{kj}^{(i-1)} - \alpha_{ki} b_{ij}^{(i)} \quad \text{for} \quad k \neq i,$$

where $i = 1, \ldots, n$; $j = 1, \ldots, n$; $k = 1, \ldots, n$ and $k \neq i$. Then it may be shown that $B^{(n)}A = D$, or $A^{-1} = D^{-1}B^{(n)}$, where $D = (d_i)$, $i = 1, \ldots, n$, is a diagonal matrix whose diagonal elements are defined

$$d_i = (\text{row } i \text{ of } B^{(i-1)})(\text{column } i \text{ of } A) = (\text{row } i \text{ of } B^{(i)})(\text{column } i \text{ of } A),$$

that is,

$$d_i = \sum_{j=1}^{n} b_{ij}^{(i)} a_{ji}.$$

2.6 PARTITIONING METHODS (BLOCK DECOMPOSITION) [11, 19, 22, 23, 27, 60, 63, 84]

2.6.1 Standard Partitioning Procedure

The inverse of a real nonsingular matrix A may be computed in terms of inverses of matrices of lower order. Let A be of order n and let $p + q = n$. Partition A into four submatrices A_1, A_2, A_3, A_4 of dimensions $p \times p, p \times q, q \times p, q \times q$, respectively:

$$A = \begin{pmatrix} A_1 & A_2 \\ A_3 & A_4 \end{pmatrix}.$$

The inverse of A may be similarly partitioned as

$$A^{-1} = \begin{pmatrix} \alpha_1 & \alpha_2 \\ \alpha_3 & \alpha_4 \end{pmatrix}.$$

We wish to find the α_i in terms of A_i. From $AA^{-1} = I$, the four matrix equations

$$A_1\alpha_1 + A_2\alpha_3 = I \qquad A_3\alpha_1 + A_4\alpha_3 = 0$$
$$A_1\alpha_2 + A_2\alpha_4 = 0 \qquad A_3\alpha_2 + A_4\alpha_4 = I$$

follow immediately and may be solved in several forms for $\alpha_1, \alpha_2, \alpha_3, \alpha_4$. The best formulation of this solution is one involving the minimum number of matrix inversions. This minimum is 2 and the solution procedure is

1. Compute A_1^{-1}.
2. Compute $\Delta = A_4 - (A_3A_1^{-1})A_2$.
3. Then $\alpha_4 = \Delta^{-1}$

$$\alpha_3 = -\Delta^{-1}(A_3A_1^{-1})$$
$$\alpha_2 = -(A_1^{-1}A_2)\,\Delta^{-1}$$
$$\alpha_1 = A_1^{-1} - (A_1^{-1}A_2)\alpha_3.$$

It follows that systems of linear equations of the form $AX = B$ may be solved as

$$X = \begin{pmatrix} X_1 \\ X_2 \end{pmatrix} = A^{-1}B = \begin{pmatrix} \alpha_1 & \alpha_2 \\ \alpha_3 & \alpha_4 \end{pmatrix}\begin{pmatrix} B_1 \\ B_2 \end{pmatrix},$$

that is,

$$X_1 = \alpha_1 B_1 + \alpha_2 B_2 \qquad X_2 = \alpha_3 B_1 + \alpha_4 B_2,$$

where B is of dimension $n \times r$, B_1 is $p \times r$, and B_2 is $q \times r$. Then X_1 is $p \times r$ and X_2 is $q \times r$.

Partitioning does not save time, but makes it possible to invert matrices too large for storage in the computer's high-speed memory. See Table 2.3 following. Partitioning and bordering (which follows) are also useful when inverses of intermediate matrices are desired.

2.6.2 Bordering Method [11, 19]

As a special case of the standard partitioning procedure given in Section 2.6.1, consider the $n \times n$ matrix A_n which results from bordering (i.e., adding a row and a column to) a matrix A_{n-1} of order $n - 1$ whose inverse A_{n-1}^{-1} is known.

Let

$$A_n = \left(\begin{array}{cccc:c} a_{11} & \cdots & a_{1,n-1} & & a_{1,n} \\ \cdot & \cdots & \cdot & & \cdot \\ \cdot & \cdots & \cdot & & \cdot \\ \cdot & \cdots & \cdot & & \cdot \\ a_{n-1,1} & \cdots & a_{n-1,n-1} & & a_{n-1,n} \\ \hline a_{n,1} & \cdots & a_{n,n-1} & & a_{n,n} \end{array}\right) = \left(\begin{array}{c:c} A_{n-1} & u_n \\ \hline v_n & a_{nn} \end{array}\right).$$

Using the results of the previous Section 2.6.1 with $p = n - 1$, $q = 1$, $p + q = n$,

$$A_n^{-1} = \left(\begin{array}{c|c} P_{n-1} & r_n \\ \hline q_n & \alpha_n^{-1} \end{array}\right),$$

where

$$\alpha_n = a_{n,n} - v_n A_{n-1}^{-1} u_n \qquad r_n = -A_{n-1}^{-1} u_n \alpha_n^{-1}$$

$$q_n = -\alpha_n^{-1} v_n A_{n-1}^{-1} \qquad P_{n-1} = A_{n-1}^{-1} - A_{n-1}^{-1} u_n q_n.$$

This procedure allows the inversion of a matrix by successive borderings by constructing in succession the inverses of the matrices

$$\begin{pmatrix} a_{11} & \cdots & a_{1i} \\ \cdot & \cdots & \cdot \\ \cdot & \cdots & \cdot \\ \cdot & \cdots & \cdot \\ a_{i1} & \cdots & a_{ii} \end{pmatrix}$$

for $i = 1, \ldots, n - 1$

TABLE 2.3
Partitioning of A19 for $t = 0.05$
using Gauss-Jordan with partial pivoting
$p + q = n$

n	20	20	20	30	30	30	30	50	50	50	50	50	50
p	0	5	10	0	5	10	15	0	5	10	15	20	25
q	20	15	10	30	25	20	15	50	45	40	35	30	25
f_r	2.7 (10^{-11})	1.5 (10^{-11})	1.8 (10^{-11})	9.3 (10^{-12})	7.9 (10^{-12})	1.0 (10^{-11})	8.0 (10^{-12})	2.0 (10^{-11})	2.0 (10^{-11})	2.2 (10^{-11})	2.2 (10^{-11})	2.3 (10^{-11})	2.3 (10^{-11})
f_e	9.3 (10^{-12})	8.7 (10^{-12})	6.7 (10^{-12})	8.3 (10^{-12})	8.8 (10^{-12})	5.6 (10^{-12})	7.2 (10^{-12})	5.1 (10^{-12})	5.6 (10^{-12})	4.2 (10^{-12})	4.2 (10^{-12})	4.6 (10^{-12})	5.1 (10^{-12})
t sec	1.233	1.500	1.583	3.800	4.350	4.633	4.900	16.367	17.683	18.383	19.350	19.983	20.917

NOTE: The matrix A19, $t = 0.05$ from Appendix C was partitioned in several ways for each of the orders $n = 20, 30, 50$. The inverse was then taken using Gauss-Jordan with partial pivoting on the CDC 1604A computer. Error measures f_r and f_e (defined in Section 5.1) and time t in seconds are given in the table for each case. Clearly partitioning does not save time or improve accuracy and should be used only when the full matrix cannot be handled in the high-speed storage of the computer.

Applied to systems of equations $A_nX_n = F_n$, where $A_{n-1}^{-1}F_{n-1}$ is known, the result is

$$\begin{pmatrix} A_{n-1} & u_n \\ v_n & a_{nn} \end{pmatrix}\begin{pmatrix} y \\ x_n \end{pmatrix} = \begin{pmatrix} F_{n-1} \\ f_n \end{pmatrix} \quad \text{or} \quad \begin{pmatrix} y \\ x_n \end{pmatrix} = \begin{pmatrix} P_{n-1} & r_n \\ q_n & \alpha_n^{-1} \end{pmatrix}\begin{pmatrix} F_{n-1} \\ f_n \end{pmatrix}$$

so that

$$y = P_{n-1}F_{n-1} + r_nf_n = (A_{n-1}^{-1}F_{n-1}) - A_{n-1}^{-1}u_nq_nF_{n-1} + r_nf_n$$
$$x_n = q_nF_{n-1} + \alpha_n^{-1}f_n = -\alpha_n^{-1}v_n(A_{n-1}^{-1}F_{n-1}) + \alpha_n^{-1}f_n.$$

2.6.3 Escalator Method of Morris [11, 31, 63]

Assume that we are given the $n \times n$ symmetric matrix A_n and the system $A_nX + a_{n+1} = 0$, where $a_{n+1} = (a_{i,n+1})$, $i = 1, \ldots, n$ is a column vector. Let y_{ik}, $i = 1, \ldots, k - 1$, be the solution of the system $A_{k-1}Y_k + a_k = 0$ where $a_k = (a_{i,k})$, $i = 1, \ldots, k - 1$. Then the numbers $y_{i,n+1} = x_i$, $i = 1, \ldots, n$ are the solutions for the complete system. The successive solution of the systems $A_{k-1}Y_k + a_k = 0$ for $k = 2, \ldots, n + 1$ is the basis for the escalator method.

Assume y_{ik} have been computed for $i < k \leq n$ and let

$$Y = \begin{pmatrix} 1 & y_{12} & y_{13} & \cdots & y_{1n} \\ & 1 & y_{23} & \cdots & y_{2n} \\ & & 1 & \cdots & y_{3n} \\ & 0 & & \cdot & \cdot \\ & & & \cdot & \cdot \\ & & & \cdot & \cdot \\ & & & & 1 \end{pmatrix}.$$

Let $C = A_nY$. Then C has zeros above the main diagonal in view of the definition of y_{ij}. The nonzero elements of C are

$$c_{ij} = \sum_{k=1}^{j-1} a_{ik}y_{kj} + a_{ij} \quad \text{for} \quad i \geq j.$$

Now $A_n = CY^{-1} = \tilde{C}DY^{-1}$ where $\tilde{C}$ is C with unit diagonal, D is a diagonal matrix, and $A_n = A_n' = (Y^{-1})'D(\tilde{C})'$. By Theorem 2, Appendix B, the factorization of A_n is unique, so that $\tilde{C}' = Y^{-1}$. Note that $D = (c_{ii})$ and

$$\tilde{C} = \left(\frac{c_{ij}}{c_{jj}}\right), \quad i > j.$$

Assume that the first k columns of Y, C, and $\tilde{C}$ have been computed. To find the $(k + 1)$th column of Y, that is, $y_{i,k+1}$ for $i \leq k$, the following recurrence formulas are derived from the relation $Y(\tilde{C})' = I$:

$$y_{i,k+1} = -\left[\sum_{j=i+1}^{k} \tilde{c}_{k+1,j}y_{ij} + \tilde{c}_{k+1,i}\right] \quad \text{for} \quad i \leq k \quad \text{and} \quad k = 2, \ldots, n + 1.$$

The connection between Morris' method and the triangularization methods is apparent, since A_n has been expressed as $A_n = CY^{-1}$ where C is upper triangular and Y^{-1} is lower triangular.

2.6.4 Inversion of Modified Matrices [19, 46, 95, 96]

2.6.4.1 Adjustment in the Inverse of a Matrix as a Result of Changing One Element in the Original Matrix

Let $A = (a_{ij})$ be the original matrix and $A^{-1} = (\alpha_{ij})$ its inverse. Let $\tilde{A} = (\tilde{a}_{ij})$ where $\tilde{a}_{ij} = a_{ij}$ except $\tilde{a}_{rs} = a_{rs} + \Delta a_{rs}$, and $\tilde{A}^{-1} = (\tilde{\alpha}_{ij})$. Then

$$\tilde{\alpha}_{ij} = \alpha_{ij} - \frac{\alpha_{ir}\alpha_{sj}\,\Delta a_{rs}}{1 + \alpha_{sr}\,\Delta a_{rs}} \quad \text{for} \quad i = 1, \ldots, n \quad \text{and} \quad j = 1, \ldots, n,$$

provided the denominator is not zero.

2.6.4.2 Adjustment in the Inverse of a Matrix as a Result of Changing One Column of the Original Matrix

Suppose $A = (a_{ij})$ differs from $\tilde{A} = (\tilde{a}_{ij})$ in the kth column. Then to compute $\tilde{A}^{-1} = (\tilde{\alpha}_{ij})$ from $A^{-1} = (\alpha_{ij})$, define

$$y_i = \sum_{r=1}^{n} \alpha_{ir}\tilde{a}_{rk} \qquad \text{for} \quad i = 1, \ldots, n,$$

$$\tilde{\alpha}_{ij} = \alpha_{ij} - y_i\tilde{\alpha}_{kj} \quad \text{for} \quad i \neq k,\, i = 1, \ldots, n \quad \text{and} \quad j = 1, \ldots, n,$$

$$\tilde{\alpha}_{kj} = \frac{\alpha_{kj}}{y_k} \qquad \text{for} \quad j = 1, \ldots, n.$$

These relationships can be used to compute any inverse matrix in n steps whenever a given matrix and its inverse are known. If A and A^{-1} are both taken as the identity matrix, then this method is equivalent to the Gaussian elimination method. Theorem 47, Appendix B may also be used to obtain the preceding results.

2.6.5 Inversion of a Complex Matrix in Terms of Inverses of Real Matrices and Solution of Systems of Complex Linear Equations [19, 23, 63, 84]

In general we wish to solve the complex system $CZ = W$, where $C = A + iB$, $Z = X + iY$ and $W = U + iV$. The matrices A, B, X, Y, U, and V are real. Then $(A + iB)(X + iY) = U + iV$, which provides the two equations $AX - BY = U$ and $BX + AY = V$. Written in matrix form,

$$\begin{pmatrix} A & -B \\ B & A \end{pmatrix}\begin{pmatrix} X \\ Y \end{pmatrix} = \begin{pmatrix} U \\ V \end{pmatrix}.$$

Hence, if A and B are $n \times n$ matrices, we can find X and Y by inverting the $2n \times 2n$ matrix

$$\begin{pmatrix} A & -B \\ B & A \end{pmatrix}.$$

However, if we consider this matrix as partitioned and similarly partition the inverse matrix as

$$\begin{pmatrix} A & -B \\ B & A \end{pmatrix}^{-1} = \begin{pmatrix} W_1 & W_2 \\ W_3 & W_4 \end{pmatrix},$$

the results of Section 2.6.1 can be used to obtain

$$\begin{aligned} \Delta &= A + BA^{-1}B, \\ W_4 &= \Delta^{-1}, \\ W_3 &= -\Delta^{-1}BA^{-1}, \\ W_2 &= A^{-1}B\Delta^{-1} = -W_3, \\ W_1 &= A^{-1} - A^{-1}B\Delta^{-1}BA^{-1} = \Delta^{-1}. \end{aligned}$$

This last result follows from Theorem 47, Appendix B.

Then

$$\begin{pmatrix} X \\ Y \end{pmatrix} = \begin{pmatrix} W_1 & -W_3 \\ W_3 & W_1 \end{pmatrix} \begin{pmatrix} U \\ V \end{pmatrix} \quad \text{and} \quad \begin{aligned} X &= W_1U - W_3V \\ Y &= W_3U + W_1V. \end{aligned}$$

In particular, when $U = I$ and $V = 0$, the inverse matrix $C^{-1} = X + iY = W_1 + iW_3$ is obtained.

Using the number of operations table in Chapter 7 for some particular method, say Gaussian elimination, an interesting comparison may be made between the number of real operations required to find C^{-1} or to solve $Cz = w$ by each of the following methods.

1. C, z, w are all real; real Gaussian elimination is used.
2. C, z, w are all complex; Gaussian elimination with complex arithmetic is used. The number of real operations needed to perform the basic complex arithmetic is indicated by the brief table below:

	Real Add or Subtract	Real Multiply	Real Divide
Complex Add or Subtract	2	0	0
Complex Multiply	2	4	0
Complex Divide	3	8	1

3. C, z, w are all complex; the partitioning method of this section is used and A^{-1} and Δ^{-1} are found by real Gaussian elimination.

4. C, z, w are all complex; the $2n \times 2n$ real system

$$\begin{pmatrix} A & -B \\ B & A \end{pmatrix} \begin{pmatrix} x \\ y \end{pmatrix} = \begin{pmatrix} u \\ v \end{pmatrix}$$

is solved by real Gaussian elimination.

Method	Number of Real Operations Required					
	To Solve $Cz = w$			To Find C^{-1}		
	$\div$	$\times$	$+$	$\div$	$\times$	$+$
I	n	$\frac{n^3}{3} + n^2 - \frac{n}{3}$	$\frac{n^3}{3} + \frac{n^2}{2} - \frac{5n}{6}$	n	$n^3 - 1$	$n^3 - 2n^2 + n$
II	n	$\frac{4n^3}{3} + 4n^2 + \frac{20n}{3}$	$\frac{4n^3}{3} + 3n^2 + \frac{2n}{3}$	n	$4n^3 + 8n - 4$	$4n^3 - 4n^2 + 5n - 2$
III	$2n$	$5n^3 + 4n^2 - 2$	$5n^3 - 2n^2$	$2n$	$5n^3 - 2$	$5n^3 - 6n^2 + 2n$
IV	$2n$	$\frac{8n^3}{3} + 4n^2 - \frac{2n}{3}$	$\frac{8n^3}{3} + 2n^2 - \frac{5n}{3}$	$2n$	$8n^3 - 1$	$8n^3 - 8n^2 + 2n$

Several conclusions may be drawn from this table:

(1) Solving $Cz = w$ or finding C^{-1} using complex Gaussian elimination on complex matrices requires about four times as much work as real Gaussian elimination performed on real matrices.

(2) It is much more efficient to use complex arithmetic as in method (II) than to use either the partitioning method (III) or the $2n \times 2n$ array method (IV) when solving complex systems of equations or inverting complex matrices.

(3) Method (IV) is more efficient than method (III) for solving $Cz = w$, whereas the reverse is true for finding C^{-1}. Neither method should be used if complex arithmetic is available on the computer, since method (II) is far more efficient.

2.6.6 Rank Annihilation [106, 107]

As a consequence of Theorem 47, Appendix B, letting A be $N \times N$ with known inverse and u and v be column vectors, the relation

$$(A + uv')^{-1} = A^{-1} - \frac{(A^{-1}u)(v'A^{-1})}{(1 + v'A^{-1}u)}$$

is obtained. An arbitrary matrix can be written as a sum of matrices of rank 1, so that by applying the above formula, any nonsingular matrix B can be inverted. Suppose the given matrix B can be written as

$$B = D + \sum_{r=1}^{N} u_r v_r',$$

where D is diagonal. Then define $C^{(1)} = D^{-1}$ and

$$C^{(n+1)} = \left[\sum_{r=1}^{n} u_r v_r' + D\right]^{-1} \quad \text{for} \quad n = 1, \ldots, N.$$

Thus $C^{N+1} = B^{-1}$. Also, $C^{(n+1)}$ may be written recursively as

$$C^{(n+1)} = C^{(n)} - \frac{(C^{(n)}u_n)(v_n'C^{(n)})}{1 + v_n'C^{(n)}u_n},$$

using the relation above for the inverse. Now choose a decomposition of B so as to minimize the labor in obtaining B^{-1}. One efficient method is the following:

Assume $B = (b_{ij})$ is $N \times N$, nonsingular, and that $b_{11} \neq 1$. Let $A^{(1)} = B - I$ and define

$$A^{(n+1)} = (a_{ij}^{(n+1)}) = \left(a_{ij}^{(n)} - \left(\frac{a_{in}^{(n)} a_{nj}^{(n)}}{a_{nn}^{(n)}}\right)\right)$$

for $n = 1, \ldots, N$. It can be shown that $a_{ij}^{(n)} = 0$ for $i < n$ or $j < n$ for every $n = 1, \ldots, N$, and that $A^{(N+1)} = 0$. By summing $A^{(n+1)}$ from $n = 1, \ldots, N$, B can now be written in the desired form as

$$B = (b_{ij}) = \left(\delta_{ij} + \sum_{r=1}^{N} \frac{a_{ir}^{(r)} a_{rj}^{(r)}}{a_{rr}^{(r)}}\right), \quad \text{so that} \quad D = I,$$

$$u_r = (u_{ir}) = \frac{a_{ir}^{(r)}}{a_{rr}^{(r)}} \quad \text{and} \quad v_r' = (v_{rj}) = a_{rj}^{(r)}.$$

CALCULATION PROCEDURE Given $B = (b_{ij})$, nonsingular and $b_{11} \neq 1$, compute

(1) $a_{ij}^{(1)} = b_{ij} - \delta_{ij}$ and $c_{ij}^{(1)} = \delta_{ij}$ for $i, j = 1, \ldots, N$.

For each $n = 1, \ldots, N$ compute

(2) $a_{ij}^{(n+1)} = a_{ij}^{(n)} - \dfrac{a_{in}^{(n)} a_{nj}^{(n)}}{a_{nn}^{(n)}}$ for $i, j > n$

(3) $u_i = \sum_{k=n}^{N} c_{ik}^{(n)} a_{kn}^{(n)}$ for $i = 1, \ldots, N$

(4) $v_j = \sum_{k=n}^{N} a_{nk}^{(n)} c_{kj}^{(n)}$ for $j = 1, \ldots, N$

(5) $\lambda = \sum_{j=n}^{N} v_j a_{jn}^{(n)} + a_{nn}^{(n)}$

(6) $c_{ij}^{(n+1)} = c_{ij}^{(n)} - \dfrac{u_i v_j}{\lambda}$ for $i, j = 1, 2, \ldots, N.$

Then $B^{-1} = C^{(N+1)}$.

If during the computation we should find $A^{(n)} \equiv 0$, that is, all $a_{ij}^{(n)} = 0$, then it would follow that $u_i^{(n)} = 0$ for all i, and $v_j^{(n)} = 0$ for all j, and $\lambda = 0$. The condition would be detected by a zero test on the denominators $a_{nn}^{(n)}$ and λ.

Since

$$C^{(n+1)} = \left[\sum_{r=1}^{n} u^{(r)}(v^{(r)})' + I\right]^{-1}$$

$$= \left[\sum_{r=1}^{n-1} u^{(r)}(v^{(r)})' + I + u^{(n)}(v^{(n)})'\right]^{-1}$$

and $u^{(n)} = v^{(n)} = 0$, we have $C^{(n+1)} = C^{(n)}$. It follows that $C^{(n)} = C^{(n+1)} = \cdots = C^{(N+1)} = B^{-1}$, so that the last C matrix computed, $C^{(n)}$, is the approximation to the inverse, B^{-1}. If B is symmetric, then so are A and C and $u = v$, so that the process is then twice as fast.

2.7 TRIDIAGONAL SYSTEMS [12, 22, 38]

A tridiagonal system is one of the form

$$\begin{aligned}
b_1x_1 + c_1x_2 &= d_1\\
a_2x_1 + b_2x_2 + c_2x_3 &= d_2\\
a_3x_2 + b_3x_3 + c_3x_4 &= d_3\\
&\vdots\\
a_{n-1}x_{n-2} + b_{n-1}x_{n-1} + c_{n-1}x_n &= d_{n-1}\\
a_nx_{n-1} + b_nx_n &= d_n\ .
\end{aligned}$$

Such systems arise frequently in the solution of difference equations resulting from partial differential equations.

Use an elimination procedure to reduce the system to one with only two unknowns per row. Call the new coefficients $\tilde{a}_i$, $\tilde{b}_i$, $\tilde{c}_i$, $\tilde{d}_i$. Also define

$a_1 = 0$ and $c_n = 0$. The new coefficients in terms of the old ones are

$$\tilde{a}_i = 0 \quad \text{for} \quad i = 2, \ldots, n$$

$$\tilde{b}_i = 1 \quad \text{for} \quad i = 1, \ldots, n$$

$$\tilde{c}_1 = \frac{c_1}{b_1} \quad \text{and} \quad \tilde{d}_1 = \frac{d_1}{b_1}$$

$$\tilde{c}_{i+1} = \frac{c_{i+1}}{b_{i+1} - a_{i+1}\tilde{c}_i}$$

$$\tilde{d}_{i+1} = \frac{d_{i+1} - a_{i+1}\tilde{d}_i}{b_{i+1} - a_{i+1}\tilde{c}_i} \quad \text{for} \quad i = 1, \ldots, n-1.$$

The solution obtained by back substitution is

$$x_n = \tilde{d}_n$$

$$x_i = \tilde{d}_i - \tilde{c}_i x_{i+1} \quad \text{for} \quad i = n-1, \ldots, 1.$$

This algorithm for tridiagonal systems involves a number of arithmetic operations which is proportional to n rather than to $\frac{1}{3}n^3$, as required by Gaussian elimination on a general matrix. The algorithm appears to have been presented first by Thomas (1949), then used in connection with parabolic partial differential equations by Bruce, Peaceman, Rachford, and Rice.

Similar methods frequently work effectively for larger band or codiagonal matrices where a Gaussian–elimination procedure on submatrices produces a suitable algorithm. Examples by Cornock (1954) and by Meyer and Hollingsworth (1957) are given in [22] pp. 99–104.

S. Schechter [93] treats systems $Qx = b$, where Q is quasi-tridiagonal, by reducing Q to the product of an upper and a lower triangular matrix each partitioned in the same manner as Q. A quasi-tridiagonal matrix Q is one of the form

$$Q = \begin{bmatrix} M_1 & E_1 & 0 & \cdots & 0 \\ D_2 & M_2 & E_2 & \cdots & 0 \\ 0 & D_3 & M_3 & E_3 & 0 \\ 0 & 0 & \cdot & \cdot & 0 \\ 0 & 0 & D_{q-1} & M_{q-1} & E_{q-1} \\ 0 & 0 & \cdot & D_q & M_q \end{bmatrix},$$

where D_k, M_k, and E_k have the same number of rows; E_k, M_{k+1}, and D_{k+2} have the same number of columns.

3

ITERATIVE METHODS

3.1 GENERAL ITERATIVE PROCEDURES [7, 11, 12, 13, 19, 20, 27, 33, 38, 40, 50, 67]

An iterative method is a rule for operating on previous approximate solutions to obtain an improved solution. Iterative methods are preferred for solving large sparse systems (such as those arising from finite-difference approximations for partial differential equations) because they can take advantage of zeros in the matrix and tend to be self-correcting and hence tend to minimize roundoff error. Such methods are particularly good for almost-diagonal or dominant-diagonal systems in which convergence is rapid. They are not so useful for finding inverse matrices or for several systems of linear equations, since a separate process has to be carried out for each column of constant terms in the B matrix. Therefore only the single system of linear equations, written $Ax = b$, is considered here. A further drawback to iterative methods is the possibility of slow or irregular convergence. The latter occurs when A has complex eigenvalues. Slow convergence may often be remedied by use of one of the acceleration procedures described in Section 3.4. Chapter 8 gives advantages and disadvantages of iterative methods and compares them with direct methods.

Given the linear system $Ax = b$, define an *iteration of degree m* as a function of the form

$$x^{(k+1)} = F_k(A, b, x^{(k)}, x^{(k-1)}, \ldots, x^{(k-m+1)}),$$

where $x^{(k)}$ is the n-component vector obtained from the kth iteration. It is desired that $x^{(k)} \to h = A^{-1}b$ as $k \to \infty$. Usually the degree of m is kept small because of storage requirements. The iteration is called *stationary* if $F_k = F$ for all k, that is, F_k is independent of k. The process is *linear* if F_k is a linear function of $x^{(k)}, \ldots, x^{(k-m+1)}$. A *Wittmeyer process* is a nonstationary linear iteration of degree $m = 1$.

Assume hereafter that the iteration is of degree $m = 1$ (unless it is specifically stated otherwise), so that

$$x^{(k+1)} = F_k(A, b, x^{(k)}).$$

A first-degree iteration is commonly written in one of the forms

$$x^{(k+1)} = x^{(k)} + C^{(k)}p^{(k)},$$

or,

$$x^{(k+1)} = T^{(k)}x^{(k)} + q^{(k)},$$

where, in general, the matrices $C^{(k)}$ and $T^{(k)}$, and the vectors $p^{(k)}$ and $q^{(k)}$ are functions of k, A, b, and $x^{(k)}$. It is assumed that $C^{(k)}$ and $T^{(k)}$ are nonsingular and that $C^{(k)} \neq A^{-1}$.

To be of any use, the above representations must leave the *solution vector*, $h = A^{-1}b$, invariant, that is, if $x^{(k)} = A^{-1}b$, then we should obtain $x^{(k+1)} = A^{-1}b$. Hence it is required that $C^{(k)}p^{(k)} = 0$ when $x^{(k)} = A^{-1}b$ and that $q^{(k)} = (I - T^{(k)})A^{-1}b$.

Define now the kth *residual vector*, $r^{(k)} = b - Ax^{(k)}$, and the kth *error vector*, $e^{(k)} = h - x^{(k)}$, where h is the solution vector. Note that $r^{(k)} = Ae^{(k)}$. An iteration converges if and only if $\|e^{(k)}\| \to 0$ as $k \to \infty$. Because h is unknown, however, it is usually either the length of the residual vector $\|r^{(k)}\|$, or the length $\|x^{(k)} - x^{(k-1)}\|$ that is used in a convergence test.

It is convenient in the first representation of the first-degree iteration to let $p^{(k)} = r^{(k)}$. Note the requirement $C^{(k)}p^{(k)} = 0$ is then satisfied when $x^{(k)} = A^{-1}b$. The iteration may then be written

$$x^{(k+1)} = x^{(k)} + C^{(k)}r^{(k)} = (I - C^{(k)}A)x^{(k)} + C^{(k)}b.$$

In the event that the iterations are linear $C^{(k)}$ and $T^{(k)}$ do not depend on $x^{(k)}$, but only on k, A, and b. For the first-degree iterations

$$x^{(k+1)} = (I - C^{(k)}A)x^{(k)} + C^{(k)}b,$$

or

$$x^{(k+1)} = T^{(k)}x^{(k)} + (I - T^{(k)})A^{-1}b,$$

which are satisfied by $x^{(k)} = x^{(k+1)} = A^{-1}b$, we have

$$A^{-1}b = (I - C^{(k)}A)A^{-1}b + C^{(k)}b,$$

so that subtracting the quantity

$$x^{(k+1)} = (I - C^{(k)}A)x^{(k)} + C^{(k)}b$$

gives

$$(A^{-1}b - x^{(k+1)}) = (I - C^{(k)}A)(A^{-1}b - x^{(k)}),$$

or

$$e^{(k+1)} = (I - C^{(k)}A)e^{(k)}.$$

Similarly, for the second representation,

$$e^{(k+1)} = T^{(k)}e^{(k)}.$$

Hence

$$e^{(k+1)} = \prod_{i=0}^{k} T^{(i)}e^{(0)} \quad \text{or} \quad e^{(k+1)} = \prod_{i=0}^{k} (I - C^{(i)}A)e^{(0)}.$$

The matrices $T^{(k)}$ and $(I - C^{(k)}A)$ are called *iteration or error-reducing matrices* and the respective iterations converge if and only if

$$M_{k+1} = \prod_{i=0}^{k} T^{(i)} \to 0 \quad \text{or} \quad M_{k+1} = \prod_{i=0}^{k}(I - C^{(i)}A) \to 0 \quad \text{as} \quad k \to \infty.$$

If all eigenvalues of M_{k+1} are less than 1 in absolute value then the iterations converge. Let $\lambda(M_k)$ be the *spectral radius* of M_k, that is, the magnitude of the eigenvalue of M_k of largest magnitude. Then $R(M_k) = -(1/k)\log[\lambda(M_k)]$ is defined as the *average rate of convergence* of the iteration for fixed k.

Note that since $r^{(k)} = Ae^{(k)}$ or $e^{(k)} = A^{-1}r^{(k)}$, we have

$$r^{(k+1)} = (AT^{(k)}A^{-1})r^{(k)}$$

and

$$r^{(k+1)} = (I - AC^{(k)})r^{(k)}.$$

By Theorem 25, Appendix B, $AC^{(k)}$ and $C^{(k)}A$ have the same eigenvalues, and by Theorem 44, Appendix B, $T^{(k)}$ and $AT^{(k)}A^{-1}$ have the same eigenvalues. Hence the convergence criteria for $e^{(k)} \to 0$ and $r^{(k)} \to 0$ in terms of eigenvalues of the iteration matrices are the same.

The situation is somewhat simplified whenever the iteration is stationary. In this case $C^{(k)}$ and $T^{(k)}$ do not depend on k at all and may be written C and T, respectively. Then $(I - CA)$ and T are the iteration matrices and

$$e^{(k+1)} = T^{k}e^{(0)} = (I - CA)^{k}e^{(0)}.$$

Consider the first-order stationary linear iteration

$$x^{(k+1)} = Tx^{(k)} + q$$

with $q = (I - T)A^{-1}b$ and solution vector $h = (I - T)^{-1}q$. Define the residuals $\tilde{r}^{(k)} = q - (I - T)x^{(k)}$. Then the following relations hold:

$$\tilde{r}^{(k)} = (I - T)A^{-1}r^{(k)}, \qquad \tilde{r}^{(k)} = (I - T)e^{(k)},$$

$$\tilde{r}^{(k)} = x^{(k+1)} - x^{(k)}, \qquad \tilde{r}^{(k+1)} = T\tilde{r}^{(k)} = T^{k+1}\tilde{r}^{(0)}.$$

Hence

$$\begin{aligned} x^{(k+1)} &= x^{(k)} + \tilde{r}^{(k)} \\ &= x^{(0)} + \tilde{r}^{(0)} + \tilde{r}^{(1)} + \cdots + \tilde{r}^{(k)} \\ &= x^{(0)} + (I + T + T^2 + \cdots + T^k)\tilde{r}^{(0)}. \end{aligned}$$

In the stationary case also

$$M_k = T^k = (I - CA)^k \to 0 \quad \text{as} \quad k \to \infty$$

defines convergence, which occurs if and only if all eigenvalues of T or of $(I - CA)$ are less than 1 in absolute value (Theorem 30, Appendix B). The *asymptotic rate of convergence* is then defined by

$$R(M_k) = -(1/k) \log [\lambda(M_k)] = -\log [\lambda(T)] = R(T).$$

A measure of the effectiveness of an iteration is the ratio of the Euclidean norm of the kth error vector $e^{(k)}$ to the Euclidean norm of the initial error vector $e^{(0)}$. The number of iterations k necessary to reduce an initial error by a prescribed factor ϵ may be estimated (see Theorem 65, Appendix B) from the following relation:

$$\max_{e^{(0)} \neq 0} \frac{\|e^{(k)}\|_E}{\|e^{(0)}\|_E} = [\lambda(M_k' M_k)]^{1/2} < \epsilon,$$

or, in the case where the iteration is stationary and T is symmetric,

$$\max_{e^{(0)} \neq 0} \frac{\|e^{(k)}\|_E}{\|e^{(0)}\|_E} = [\lambda(T)]^k < \epsilon,$$

so that the required k satisfies

$$k > \frac{-\log \epsilon}{-\log \lambda(T)} = \frac{-\log \epsilon}{R}.$$

A particular iteration (be it linear or nonlinear, stationary or nonstationary) depends on the choice of $C^{(k)}$ and $p^{(k)}$. Often this choice is made so as to minimize some measure of error, $\phi(x)$. Three such measures are

$$\phi_1(x) = (h - x, h - x) = (e, e) = (A^{-1}r, A^{-1}r) = (r, (AA')^{-1}r)$$
$$\phi_2(x) = (r, r) = (h - x, A'A(h - x)) = (e, A'Ae)$$
$$\phi_3(x) = (h - x, A(h - x)) = (e, Ae) = (e, r) = (A^{-1}r, r) = (r, (A')^{-1}r).$$

Note that by Theorem 53, Appendix B, $(Ax, y) = (x, A'y)$ in the real case, and $(Ax, y) = (x, (A^*)'y)$ for the complex. Also $A(h - x) = Ah - Ax = b - Ax = r$, or $e = A^{-1}r$.

$\phi_1(x)$ is the square of the length of the error vector and $\phi_2(x)$ is the square of the length of the residual vector. $\phi_3(x)$ should be used only when A is symmetric positive definite, otherwise its minimum might be negative.

Having classified an iteration according to its degree, according to whether it is linear or nonlinear and according to whether it is stationary or nonstationary, we now consider a further distinction. An *explicit iteration* or *point iteration* or *single-step iteration* is one for which each coordinate $x_i^{(k+1)}$ of the new iteration vector $x^{(k+1)}$ can be determined by itself without the necessity of determining a group of other components $x_j^{(k+1)}$ simultaneously. An *implicit*, *block*, or *group iteration*, on the other hand, divides the equations into groups, then solves the subsystem of equations belonging to a group for the corresponding unknowns $x_i^{(k+1)}$ using approximate values for the other unknowns $x_j^{(k+1)}$. The grouping should be such that the subsystems are easy to solve, for example diagonal, triangular, or tri-diagonal systems of equations.

The iterative methods that follow are, with one exception, of degree 1 and are classified first with respect to linearity, next according to whether they are point or block iterations, then according to whether or not they are stationary. There is nothing sacred in this hierarchy; other orderings would serve equally well.

Write the general iteration of degree 1 in the form

$$x^{(k+1)} = x^{(k)} + C^{(k)}p^{(k)}.$$

In the methods that follow, choices for the matrix $C^{(k)}$ include a scalar $\alpha^{(k)}$, a diagonal matrix, and a constant matrix. Possible choices for the vector $p^{(k)}$ include the gradient direction, the residual direction, the directions of the coordinate axes, the directions A-orthogonal to the coordinate axes, and the directions A-orthogonal to the gradient. Many of the iterations can be expressed in several ways using the above form.

3.2 NONLINEAR ITERATIONS

3.2.1 Gradient Methods

3.2.1.1 Preliminary Discussion

Assume that the matrix S is symmetric and positive definite. Then any of the three error measures can be written in the form $\phi(x) = (r, Sr)$. For $\phi_1(x)$, $S = (AA')^{-1}$; for $\phi_2(x)$, $S = I$; for $\phi_3(x)$, $S = (A')^{-1}$.

Take the case now where $C^{(k)} = \alpha^{(k)}$, $\alpha^{(k)}$ a scalar. The iteration becomes $x^{(k+1)} = x^{(k)} + \alpha^{(k)}p^{(k)}$. Hence from the point $x^{(k)}$, the approximation proceeds in the direction $p^{(k)}$ for a distance $\alpha^{(k)}$ along the surface $\phi(x)$, attempting to choose $\alpha^{(k)}$ and $p^{(k)}$ at each step so as to approach the minimum point of $\phi(x)$. The best choice for $p^{(k)}$ would, of course, be the direction of the error vector $h - x$, but this vector is unfortunately not known.

The *gradient* of $\phi(x)$ at the point $x^{(k)}$ is a vector with n components

$$\left(\frac{\partial \phi}{\partial x_i}\right)_{x^{(k)}}, \qquad i = 1, \ldots, n.$$

At $x = h$ the gradient will be zero. The gradient is a vector in $(x_1, \ldots, x_n)$ space normal to the contour surface (not the surface) through $x^{(k)}$; it is the direction of steepest descent at $x^{(k)}$. A possible choice for $p^{(k)}$ is the gradient direction of $\phi(x)$ at $x^{(k)}$.

For the general error measure $\phi(x) = (r, Sr)$ with S symmetric and positive definite, the gradient direction will be computed as well as the optimum value of α required to minimize $\phi(x + \alpha p)$ for arbitrary x and p.

(a) *Gradient of* $\phi(x)$.

$$\phi(x) = (r, Sr) = (Sr, r) = \sum_{i=1}^{n}(Sr)_i(r)_i = \sum_{i=1}^{n}\sum_{j=1}^{n}(s_{ij}r_j)r_i,$$

$$\frac{\partial\phi}{\partial x_k} = \sum_{i=1}^{n}\sum_{j=1}^{n}\left[s_{ij}\left(r_i\frac{\partial r_j}{\partial x_k} + r_j\frac{\partial r_i}{\partial x_k}\right)\right].$$

Now S is symmetric, so $s_{ij} = s_{ji}$ and

$$\frac{\partial\phi}{\partial x_k} = 2\sum_{i=1}^{n}\sum_{j=1}^{n}\left(s_{ij}r_j\frac{\partial r_i}{\partial x_k}\right).$$

Since

$$r_i = (b - Ax)_i = b_i - \sum_{j=1}^{n}a_{ij}x_j,$$

it follows that $\partial r_i/\partial x_k = -a_{ik}$. Then

$$\frac{\partial\phi}{\partial x_k} = -2\sum_{i=1}^{n}\sum_{j=1}^{n}s_{ij}r_ja_{ik} = -2\sum_{i=1}^{n}a_{ik}\sum_{j=1}^{n}s_{ij}r_j = -2\sum_{i=1}^{n}a_{ik}(Sr)_i.$$

Therefore gradient $\phi(x) = -2A'Sr$. Substitute for S to obtain

$$\text{gradient } \phi_1(x) = -2A^{-1}r,$$

$$\text{gradient } \phi_2(x) = -2A'r,$$

$$\text{gradient } \phi_3(x) = -2r.$$

(b) *Optimum value of* α *required to minimize* $\phi(x + \alpha p)$ *for arbitrary* x *and* p.

$$\phi(x) = (r, Sr) = (b - Ax, Sb - SAx).$$

$$\begin{aligned}\phi(x + \alpha p) &= (b - AX - \alpha Ap, Sb - SAx - \alpha SAp)\\ &= (r - \alpha Ap, Sr - \alpha SAp)\\ &= (r, Sr) - \alpha(r, SAp) - \alpha(Ap, Sr) + \alpha^2(Ap, SAp).\end{aligned}$$

Now

$$(Ap, Sr) = (S'Ap, r) = (SAp, r) = (r, SAp),$$

So

$$\phi(x + \alpha p) = (r, Sr) - 2\alpha(r, SAp) + \alpha^2(Ap, SAp).$$

Then

$$\frac{\partial \phi}{\partial \alpha} = -2(r, SAp) + 2\alpha(Ap, SAp) = 0,$$

so that

$$\alpha = \frac{(r, SAp)}{(Ap, SAp)}.$$

Substitute for S to obtain the α's corresponding to ϕ_1, ϕ_2, ϕ_3:

$$\text{for} \quad \phi_1(x), \qquad \alpha = \frac{(r, (A')^{-1}p)}{(Ap, (A')^{-1}p)} = \frac{(A^{-1}r, p)}{(p, p)}$$

$$\text{for} \quad \phi_2(x), \qquad \alpha = \frac{(r, Ap)}{(Ap, Ap)}$$

$$\text{for} \quad \phi_3(x), \qquad \alpha = \frac{(A^{-1}r, Ap)}{(p, Ap)} = \frac{(A'A^{-1}r, p)}{(p, Ap)} = \frac{(r, p)}{(p, Ap)}.$$

3.2.1.2 Gradient Methods (Steepest Descent) [8, 13, 27, 35, 60, 67, 97]

To solve the system $Ax = b$ start with a trial point $x^{(0)}$ in n-space and move in the direction $p^{(k)}$ to the new approximation $x^{(1)}$, adding a correction of $\alpha^{(k)}p^{(k)}$ to $x^{(0)}$. Gradient methods are characterized by the fact that $p^{(k)}$ is chosen to be the gradient direction. If, in addition, $\alpha^{(k)}$ is taken as the optimum α required to minimize ϕ for arbitrary x and p, the optimum-gradient or steepest-descent methods are obtained. These methods are stationary, explicit, and nonlinear. For each of the three error measures, the summary below gives $p^{(k)}$ and $\alpha^{(k)}$, $\tilde{p}^{(k)}$ and $C^{(k)}$ for the iteration,

$$x^{(k+1)} = x^{(k)} + \alpha^{(k)}p^{(k)} = x^{(k)} + C^{(k)}\tilde{p}^{(k)}.$$

(1) $\phi_1(x) = (h - x, h - x)$.

$$p^{(k)} = A^{-1}r^{(k)}, \qquad \alpha^{(k)} = \frac{(A^{-1}r^{(k)}, p^{(k)})}{(p^{(k)}, p^{(k)})} = \frac{\|A^{-1}r^{(k)}\|^2}{\|p^{(k)}\|^2} = 1$$

$$\tilde{p}^{(k)} = r^{(k)}, \qquad C^{(k)} = \alpha^{(k)}A^{-1} = A^{-1}$$

(2) $\phi_2(x) = (r, r)$: Method of Cauchy.

$$p^{(k)} = A'r^{(k)}, \qquad \alpha^{(k)} = \frac{(r^{(k)}, Ap^{(k)})}{(Ap^{(k)}, Ap^{(k)})} = \frac{(r^{(k)}, AA'r^{(k)})}{\|AA'r^{(k)}\|^2}$$

$$\tilde{p}^{(k)} = r^{(k)}, \; C^{(k)} = \alpha^{(k)}A'.$$

(3) $\phi_3(x) = (h - x, A(h - x))$, **$A$ symmetric and positive definite.**

$$p^{(k)} = r^{(k)}, \qquad \alpha^{(k)} = \frac{(p^{(k)}, r^{(k)})}{(p^{(k)}, Ap^{(k)})} = \frac{(r^{(k)}, r^{(k)})}{(r^{(k)}, Ar^{(k)})} = \frac{\|r^{(k)}\|^2}{(r^{(k)}, Ar^{(k)})}$$

$$\tilde{p}^{(k)} = r^{(k)}, \qquad C^{(k)} = \alpha^{(k)} I.$$

(4) $\phi_1(x) = (h - x, h - x)$: **Method of Householder.**

$$p^{(k)} = A' r^{(k)}, \qquad \alpha^{(k)} = \frac{\|r^{(k)}\|^2}{\|A' r^{(k)}\|^2}$$

$$\tilde{p}^{(k)} = r^{(k)}, \qquad C^{(k)} = \alpha^{(k)} A'.$$

Note first that (1) is of no use because A^{-1} is not known. Householder has suggested (4) for use with the measure $\phi_1(x)$. Method (3) is the familiar form of steepest descent. Note also that any of the four above variations may be written in the form

$$x^{(k+1)} = x^{(k)} + C^{(k)} r^{(k)} = (I - C^{(k)} A) x^{(k)} + C^{(k)} b$$

with error vector

$$e^{(k+1)} = (I - C^{(k)} A) e^{(k)} = \prod_{i=0}^{k} (I - C^{(i)} A) e^{(0)} = M_{k+1} e^{(0)}.$$

The matrix $(I - C^{(k)} A)$ is called an *iteration matrix*, and the process converges if $M_{k+1} \to 0$ as $k \to \infty$. By Theorem 30, Appendix B, this occurs if every eigenvalue of M_{k+1} is less than 1 in absolute value.

Marcus [27], p. 18, gives the optimum α and convergence criteria (in terms of eigenvalues) for the complex case.

Several authors using method (3) above have described the method as good for well-conditioned systems, but have found that in ill-conditioned examples it tends to "overshoot." In this case the use of the value 0.9α is suggested to hasten convergence, (see [8, 97]). Convergence in any case is only linear, so that gradient methods are not particularly fast unless an acceleration procedure is applied.

Stein [97], using the iteration in the form $x^{(k+1)} = x^{(k)} + \beta^{(k)} \alpha^{(k)} p^{(k)}$, where $\beta^{(k)}$ is an arbitrary number, found that $\beta^{(k)} < 1$ gives better convergence than $\beta^{(k)} = 1$, the optimum-gradient method, and that $\beta^{(k)} > 1$ gives worse convergence than for $\beta^{(k)} = 1$. He observed that a large number of gradient methods are self-accelerating, that is, irregularly show large increases in the rate of convergence without any modification in the routine.

3.2.2 Conjugate-Direction Methods [8, 47, 72, 75]

Conjugate-direction methods are based on generation of a set of A-orthogonal vectors and then minimizing successively in the direction of each of them. A set of vectors $\{p^{(k)}\}$, $k = 0, \ldots, n-1$ is chosen so as to be A-conjugate (or A-orthogonal), that is, so that $(Ap^{(i)}, p^{(j)}) = 0$ for $i \neq j$. If A is positive definite, then $(Ap^{(i)}, p^{(i)}) > 0$. In this case the vectors $p^{(k)}$ are linearly independent and such that any solution h of the linear system $Ax = b$ can be written (by Theorem 70, Appendix B) in the form

$$h = \sum_{i=0}^{n-1} \alpha_i p^{(i)}.$$

Since the $p^{(k)}$ are A-orthogonal, it follows that

$$(Ah, p^{(k)}) = (b, p^{(k)}) = \sum_{i=0}^{n-1} \alpha_i (Ap^{(i)}, p^{(k)}) = \alpha_k (Ap^{(k)}, p^{(k)}),$$

so that

$$\alpha_k = \frac{(b, p^{(k)})}{(Ap^{(k)}, p^{(k)})}.$$

Once the vectors $p^{(k)}$ are determined, the solution h is obtained by an n-step iteration of the form

$$x^{(i+1)} = x^{(i)} + \alpha_i p^{(i)}, \qquad i = 0, \ldots, n-1$$

$$\alpha_i = \frac{(b, p^{(i)})}{(Ap^{(i)}, p^{(i)})}, \qquad i, = 0 \ldots, n-1$$

and beginning with initial vector $x^{(0)} = 0$. Then $x^{(n)} = h$.

To determine a set of A-orthogonal vectors $y^{(k)}$ from a given set of n linearly independent vectors $v^{(k)}$, use a Gram–Schmidt orthogonalization procedure summarized as follows:

$$y^{(0)} = v^{(0)}$$

$$y^{(k+1)} = v^{(k+1)} + \sum_{i=0}^{k} C_i^{(k+1)} y^{(i)} \quad \text{for} \quad k = 0, \ldots, n-2.$$

Assume that $y^{(0)}, \ldots, y^{(k)}$ have been found and are A-orthogonal. Then in order that

$$(Ay^{(k+1)}, y^{(r)}) = (Av^{(k+1)}, y^{(r)}) + \sum_{i=0}^{k} C_i^{(k+1)} (Ay^{(i)}, y^{(r)}) = 0$$

for all $r \leq k$, define

$$C_r^{(k+1)} = -\frac{(Av^{(k+1)}, y^{(r)})}{(Ay^{(r)}, y^{(r)})}.$$

Hence all $y^{(0)}, \ldots, y^{(n-1)}$ are defined once $v^{(0)}, \ldots, v^{(n-1)}$ have been specified.

The process may also be formulated in the following manner:

$$y^{(k+1)} = y^{(k)} + \sum_{i=0}^{k-1} C_i^{(k+1)} y^{(i)} + C_k^{(k+1)} v^{(k+1)}.$$

Assume that $y^{(0)}, \ldots, y^{(k)}$ have been found and are A-orthogonal. Then determine $C_k^{(k+1)}$ so that for $r \leq k$,

$$(Ay^{(k+1)}, y^{(r)}) = (Ay^{(k)}, y^{(r)}) + \sum_{i=0}^{k-1} C_i^{(k+1)}(Ay^{(i)}, y^{(r)}) + C_k^{(k+1)}(Av^{(k+1)}, y^{(r)}) = 0.$$

Two cases result:

(1) $r = k$: $$C_k^{(k+1)} = \frac{-(Ay^{(k)}, y^{(k)})}{(Av^{(k+1)}, y^{(k)})}$$

(2) $r < k$: $$C_r^{(k+1)} = \frac{-C_k^{(k+1)}(Av^{(k+1)}, y^{(r)})}{(Ay^{(r)}, y^{(r)})}.$$

Note that in either representation, if the matrix A is taken as I, $y^{(k)}$ is orthogonal to $v^{(k)}$. Note also that since $v^{(k)}$ is a linear combination of $y^{(0)}, \ldots, y^{(k)}$,

$$(v^{(k)}, Ay^{(i)}) = 0 \quad \text{for} \quad i > k.$$

Consider again the error measure $\phi_3(x) = (h - x, A(h - x))$, assuming A symmetric and positive definite. Its gradient direction was grad $\phi_3(x) = -2r$, and the optimum α for minimizing $\phi_3(x + \alpha p)$ was $\alpha = (p, r)/(p, Ap)$, where $r = b - Ax$ is the residual. When the $p^{(k)}$ are A-orthogonal, we may write (for $x^{(0)} = 0$)

$$(p^{(i)}, r^{(i)}) = (p^{(i)}, b - Ax^{(i)}) = (p^{(i)}, b) - \sum_{k=0}^{i-1} \alpha_k (p^{(i)}, Ap^{(k)}) = (p^{(i)}, b).$$

Then

$$\alpha_i = \frac{(p^{(i)}, r^{(i)})}{(p^{(i)}, Ap^{(i)})} = \frac{(p^{(i)}, b)}{(p^{(i)}, Ap^{(i)})}$$

so that the scalars α_i previously defined are actually optimal for the error measure $\phi_3(x)$.

Analogous results may be obtained for the measure $\phi_2(x) = (r, r) = (h - x, A'A(h - x))$ using vectors $p^{(k)}$ which are $(A'A)$-orthogonal, as well as for the measure $\phi_1(x) = (h - x, h - x)$ using vectors $p^{(k)}$ which are orthogonal. For either measure, define $x^{(i+1)} = x^{(i)} + \alpha_i p^{(i)}$ where for the measure $\phi_2(x)$,

$$\alpha_i = \frac{(Ap^{(i)}, r^{(i)})}{\|Ap^{(i)}\|^2} = \frac{(Ap^{(i)}, b)}{\|Ap^{(i)}\|^2} = \frac{(p^{(i)}, A'b)}{\|Ap^{(i)}\|^2}$$

and for the measure $\phi_1(x)$,

$$\alpha_i = \frac{(p^{(i)}, A^{-1}r^{(i)})}{\|p^{(i)}\|^2} = \frac{(p^{(i)}, A^{-1}b)}{\|p^{(i)}\|^2} = \frac{((A')^{-1}p^{(i)}, r^{(i)})}{\|p^{(i)}\|^2} .$$

Return to the original measure $\phi_3(x)$ and the iteration

$$x^{(i+1)} = x^{(i)} + \alpha_i p^{(i)}$$

with

$$\alpha_i = \frac{(p^{(i)}, r^{(i)})}{(p^{(i)}, Ap^{(i)})}$$

and

$$r^{(i)} = b - Ax^{(i)}.$$

From

$$r^{(i)} = b - Ax^{(i)} = b - A(x^{(i-1)} + \alpha_{i-1}p^{(i-1)}) = (b - Ax^{(i-1)}) - \alpha_{i-1}Ap^{(i-1)}$$

the recursion

$$r^{(i)} = r^{(i-1)} - \alpha_{i-1}Ap^{(i-1)}$$

for the residual is derived. Also, from the preceding relations and from the fact that $(Ap^{(i)}, p^{(j)}) = 0$ for $i \neq j$, it may be shown that

$$(p^{(i)}, r^{(j)}) = 0 \quad \text{for} \quad i < j$$

$$(p^{(i)}, r^{(i)}) = (p^{(i)}, r^{(i-1)}) = \cdots = (p^{(i)}, r^{(0)}).$$

Geometrically the situation is that of finding the center $x^{(k)}$ of a $(k - 1)$-dimensional ellipsoid which results from the intersection of the ellipsoid $\phi_3(x) =$ constant and the hyperplane $x = x^{(0)} + \alpha_1 p^{(1)} + \cdots + \alpha_{k-1}p^{(k-1)}$. The centers of ellipsoids of ever-increasing dimension, which are found successively, converge to the center h of $\phi_3(x) =$ constant.

The conjugate-direction method does not specify how to compute $p^{(k)}$ since it does not specify $v^{(k)}$. The case where $p^{(k)}$ is obtained by A-orthogonalization from the unit vectors leads to Gaussian elimination.

The estimates $x^{(k)}$ may be computed without computing the residuals $r^{(k)}$, provided the choice of $p^{(k)}$ is independent of $r^{(k)}$. The conjugate-direction method is considered an m-step method, $m \leq n$, where m depends on the initial approximation, $x^{(0)}$.

3.2.3 Conjugate-Gradient Methods [19, 27, 47, 57, 74, 75]

The conjugate-gradient method combines the features of the conjugate direction and gradient methods. It is a special case of the conjugate-direction method in which the vectors $p^{(k)}$ are obtained by the A-orthogonalization of the residual vectors $r^{(k)}$. A conjugate-direction method in which the residual vectors are mutually orthogonal is essentially a conjugate-gradient method.

3.2.3.1 For Symmetric Positive-Definite Matrices

Assume that the system $Ax = b$, with A symmetric, positive definite, is to be solved. Define the following iterative procedure:

Choose an initial vector $x^{(0)}$ with n components. Set $p^{(0)} = r^{(0)} = b - Ax^{(0)}$. For $i = 0, \ldots, n - 1$ compute

$$\alpha_i = \frac{\|r^{(i)}\|^2}{(p^{(i)}, Ap^{(i)})} = \frac{\sum_{j=1}^{n} (r_j^{(i)})^2}{\sum_{j=1}^{n} p_j^{(i)}(Ap^{(i)})_j}$$

or better

$$\alpha_i = \frac{(p^{(i)}, r^{(i)})}{(p^{(i)}, Ap^{(i)})} = \frac{\sum_{j=1}^{n} p_j^{(i)} r_j^{(i)}}{\sum_{j=1}^{n} p_j^{(i)}(Ap^{(i)})_j}$$

$$x^{(i+1)} = x^{(i)} + \alpha_i p^{(i)} = x^{(0)} + \sum_{k=0}^{i} \alpha_k p^{(k)}$$

$$r^{(i+1)} = r^{(i)} - \alpha_i A p^{(i)} = b - Ax^{(i+1)};$$

$$\beta_i = \frac{\|r^{(i+1)}\|^2}{\|r^{(i)}\|^2} = \frac{\sum_{j=1}^{n} (r_j^{(i+1)})^2}{\sum_{j=1}^{n} (r_j^{(i)})^2}$$

or better,

$$\beta_i = \frac{-(r^{(i+1)}, Ap^{(i)})}{(p^{(i)}, Ap^{(i)})} = \frac{-\sum_{j=1}^{n} r_j^{(i+1)}(Ap^{(i)})_j}{\sum_{j=1}^{n} p_j^{(i)}(Ap^{(i)})_j}$$

$$p^{(i+1)} = r^{(i+1)} + \beta_i p^{(i)}.$$

The values of α_i and β_i given first are simpler to compute, but the second values for α_i and β_i give better results according to Hestenes and Stiefel.

Since A is positive definite, the denominators above are all greater than 0. Once the vectors $p^{(k)}$ have been obtained, the solution of any system $Ax = \tilde{b}$ can be found from

$$x = \sum_{i=0}^{n-1} \tilde{\alpha}_i p^{(i)}$$

where

$$\tilde{\alpha}_i = \frac{(p^{(i)}, \tilde{b})}{(Ap^{(i)}, p^{(i)})}$$

and where $x^{(0)} = 0$. Also

$$A^{-1} = (\tilde{a}_{jk}) = \left(\sum_{i=0}^{n-1} \frac{p_j^{(i)} p_k^{(i)}}{(p^{(i)}, Ap^{(i)})} \right)$$

After m iterations, $m \leq n$, $x^{(m)}$ should be equal to the solution, h, of $Ax = b$, if all computations could be done with no loss of accuracy. The procedure may be continued, using $x^{(n)}$ as the initial $x^{(0)}$, until the desired accuracy is obtained. It can be shown that the error vector $h - x^{(i)}$ is decreased in length at each step; that is, $\|h - x^{(j)}\| < \|h - x^{(i)}\|$ for $j > i$. Also the measure $\phi_3(x) = (h - x, A(h - x))$ is decreased at each step. However, it is not necessarily true that $\|r^{(i)}\|$ decreases, in fact the residual $r^{(i)} = b - Ax^{(i)}$ normally oscillates and may increase. The test $\|r^{(i)}\| < \epsilon$ is usually used for convergence; a test, $|\text{denominator}| < \delta$, for a zero denominator in α_i and β_i should also be made. Careful choice of the values for ϵ and δ is important.

The preceding sequence of formulae describe the formation of two sequences of vectors by means of the Gram–Schmidt process. The residual vectors $r^{(i)}$ and the direction vectors $p^{(i)}$ are formed in the order $r^{(0)}, p^{(0)}, r^{(1)}, p^{(1)}, \ldots, r^{(n-1)}, p^{(n-1)}$. Form $r^{(0)}, \ldots, r^{(i+1)}$ by applying an orthogonalization process to $r^{(0)}, Ap^{(0)}, Ap^{(1)}, \ldots, Ap^{(i)}$; find $p^{(0)}, \ldots, p^{(i)}$ by computing a set of A-orthogonal vectors from $r^{(0)}, \ldots, r^{(i)}$. Since $r^{(n)}$ would be orthogonal to all n mutually orthogonal linearly independent vectors $r^{(i)}$, $i = 0, \ldots, n - 1$ and would be expressible as a linear combination of these n linearly independent vectors, the coefficients in this linear expression would all be zero and $r^{(n)}$ must be the zero vector.

From the statements above and from the relations given in the conjugate-direction method, the following may be shown:

$$(p^{(i)}, Ap^{(j)}) = 0 \quad \text{for} \quad i \neq j$$

$$(r^{(i)}, Ap^{(k)}) = 0 \quad \text{for} \quad i < k$$

$$(r^{(i)}, r^{(j)}) = 0 \quad \text{for} \quad i \neq j$$

$$(Ap^{(k-1)}, r^{(i)}) = 0 \quad \text{for} \quad i > k, \quad \text{or} \quad (Ap^{(k)}, r^{(i)}) = 0 \quad \text{for} \quad i > k + 1$$

$$(p^{(i)}, Ap^{(i)}) = (r^{(i)}, Ap^{(i)})$$

$$(p^{(i)}, r^{(j)}) = (p^{(i)}, r^{(i)}) = (r^{(i)}, r^{(i)}) \quad \text{for} \quad j \leq i$$

$$(p^{(i)}, r^{(j)}) = 0 \quad \text{for} \quad j > i.$$

One common difficulty with methods that use orthogonality is that roundoff errors tend to disturb the orthogonality relations. Instability may result.

The four propagation formulas below are useful for checking error buildup:

$$(r^{(i)}, r^{(i+1)}) = \beta_{i-1}\alpha_i(Ap^{(i-1)}, p^{(i)})$$

$$(Ap^{(i)}, p^{(i+1)}) = \frac{(r^{(i)}, r^{(i+1)})}{\alpha_i}$$

$$(r^{(i)}, r^{(i+1)}) = \frac{\beta_{i+1}\alpha_i}{\alpha_{i-1}}(r^{(i-1)}, r^{(i)})$$

$$(Ap^{(i)}, p^{(i+1)}) = \beta_{i-1}(Ap^{(i-1)}, p^{(i)}).$$

From the above propagation formulas and previous relations, write

$$\frac{(r^{(i)}, r^{(i+1)})}{\|r^{(i)}\|^2} = \frac{\alpha_i}{\alpha_{i-1}} \frac{(r^{(i-1)}, r^{(i)})}{\|r^{(i-1)}\|^2}$$

$$\frac{(Ap^{(i)}, p^{(i+1)})}{(Ap^{(i)}, p^{(i)})} = \frac{\alpha_i}{\alpha_{i-1}} \frac{(Ap^{(i-1)}, p^{(i)})}{(Ap^{(i-1)}, p^{(i-1)})}.$$

Hence the larger the ratio α_i/α_{i-1}, the more rapidly roundoff errors accumulate, and the greater the disturbance of the orthogonality relations.

For a given positive-definite matrix A with distinct eigenvalues, there exists an initial $r^{(0)}$ such that

$$\left(\frac{\alpha_i}{\alpha_{i-1}}\right) < 1,$$

hence such that the algorithm is stable with respect to propagation of roundoff.

Geometrically the conjugate-gradient method may be viewed as follows:

Let Π_i be the $(n - i)$-dimensional hyperplane through $x^{(i)}$ and conjugate to $p^{(0)}, p^{(1)}, \ldots, p^{(i-1)}$. It also contains the points $x^{(i+1)}, x^{(i+2)}, \ldots, x^{(n-1)}$.

Let $\phi_3(x) = \phi_3(x^{(i)})$ be the $(n - 1)$-dimensional hyperellipsoid defined by setting the error measure $\phi_3(x)$ equal to a constant $\phi_3(x^{(i)})$. The gradient of ϕ_3 at $x^{(i)}$ is $-2r^{(i)}$.

Let E_i' be the $(n - i - 1)$-dimensional hyperellipsoid formed by intersecting $\phi_3(x) = \phi_3(x^{(i)})$ and Π_i. The center of E_i' is h, the solution of the system $Ax = b$.

Let P_i be the i-dimensional hyperplane through $x^{(0)}, \ldots, x^{(i)}$.

Let E_i be the $(i - 1)$-dimensional hyperellipsoid formed by intersecting $\phi_3(x) = \phi_3(x^{(i)})$ and P_i. The center of E_i is $x^{(i)}$.

From the point $x^{(i)}$ travel in the direction $p^{(i)}$ a distance α_i to the point $x^{(i+1)}$. [α_i is chosen so as to minimize $\phi_3(x^{(i)} + \alpha_i p^{(i)})$.] The point $x^{(i+1)}$ is the midpoint of a chord of the hyperellipsoid E_i' which goes through $x^{(i)}$ and is parallel to $p^{(i)}$. The chord is normal to E_i' at $x^{(i)}$ and its direction is a scalar multiple of the projection of the gradient of $\phi_3(x)$ at $x^{(i)}$ in the plane Π_i. (β_i is chosen so that the vectors $p^{(i+1)} = r^{(i+1)} + \beta_i p^{(i)}$ are A-orthogonal.) The estimates $x^{(0)}, \ldots, x^{(m)}$ of h are distinct and $x^{(i)}$ minimizes $\phi_3(x)$ on P_i.

The iteration is continued using $x^{(i+1)}$, $\phi_3(x) = \phi_3(x^{(i+1)})$, Π_{i+1}, E'_{i+1}, etc., so that at each step the dimension of the space in which the solution h is sought is decreased by 1. After not more than n steps, the solution should (theoretically) be found.

3.2.3.2 *Variation of the Method for Symmetric Positive-Definite Matrices for Which the Residuals Also Decrease*

It was stated in Section 3.2.3.1 that, for the procedure given for solving $Ax = b$ for symmetric positive-definite matrix A, the error vector $h - x^{(i)}$ and error measure $\phi_3(x)$ decrease at each step, but that the residual $r^{(i)} = b - Ax^{(i)}$ need not decrease. A slight variation in the method plus some additional computations produce a "smoothed" set of estimates $\tilde{x}^{(0)}, \tilde{x}^{(1)}, \ldots, \tilde{x}^{(n-1)}$ for which the residual also decreases. The residual is proportional to the direction $p^{(i)}$.

Let $\tilde{x}^{(0)} = x^{(0)}$, $C_0 = 1$, $p^{(0)} = r^{(0)} = b - Ax^{(0)}$. For $i = 0, \ldots, n - 1$, compute

$$\alpha_i = \frac{\|r^{(i)}\|^2}{(p^{(i)}, Ap^{(i)})} \quad \text{or} \quad \alpha_i = \frac{(p^{(i)}, r^{(i)})}{(p^{(i)}, Ap^{(i)})}$$

$$x^{(i+1)} = x^{(i)} + \alpha_i p^{(i)}$$

$$r^{(i+1)} = r^{(i)} - \alpha_i A p^{(i)} = b - Ax^{(i+1)}$$

$$\beta_i = \frac{\|r^{(i+1)}\|^2}{\|r^{(i)}\|^2} \quad \text{or} \quad \beta_i = -\frac{(r^{(i+1)}, Ap^{(i)})}{(p^{(i)}, Ap^{(i)})}$$

$$C_{i+1} = 1 + \beta_i C_i$$

$$\tilde{x}^{(i+1)} = \frac{x^{(i+1)} + \beta_i C_i \tilde{x}^{(i)}}{C_{i+1}}$$

and

$$p^{(i+1)} = C_{i+1}(b - A\tilde{x}^{(i+1)}).$$

The following relations hold:

$$C_i = \|r^{(i)}\|^2 \sum_{j=0}^{i} \frac{1}{\|r^{(j)}\|^2} = \frac{\|p^{(i)}\|^2}{\|r^{(i)}\|^2}$$

$$\tilde{x}^{(i)} = \frac{\|r^{(i)}\|^2}{C_i} \sum_{j=0}^{i} \frac{x^{(j)}}{\|r^{(j)}\|^2}$$

$$\tilde{r}^{(i)} = b - A\tilde{x}^{(i)} = \frac{p^{(i)}}{C_i} = \frac{\|r^{(i)}\|^2}{C_i} \sum_{j=0}^{i} \frac{r^{(j)}}{\|r^{(j)}\|^2}.$$

Note that in the relations for $\tilde{x}^{(i)}$ and $\tilde{r}^{(i)}$ the sum of the coefficients of the $x^{(j)}$ is 1, and the sum of the coefficients of the $r^{(j)}$ is 1.

The residuals $\|\tilde{r}^{(i)}\|$ decrease monotonically as i increases:

$$\|\tilde{r}^{(i-1)}\|^2 - \|\tilde{r}^{(i)}\|^2 = \frac{\|\tilde{r}^{(i-1)}\|^2}{C_i},$$

and the error measure $\phi_3(\tilde{x}^{(i)})$ decreases monotonically as i increases. The error vector $h - x^{(i)}$ is shorter than the error vector $h - \tilde{x}^{(i)}$, which in turn is shorter than $h - x^{(i-1)}$. Also $\phi_3(x^{(i)}) < \phi_3(\tilde{x}^{(i)}) < \phi_3(\tilde{x}^{(i-1)})$.

3.2.3.3 *Generalization of the Method to an Arbitrary Nonsingular Matrix*

The system $Ax = b$ is equivalent to the system $(A'A)x = (A'b)$, and $(A'A)$ is both symmetric and positive definite if A is nonsingular. Thus the methods in Sections 3.2.3.1 and 3.2.3.2 can be applied to $(A'A)$ in place of A and $A'b$ in place of b. However, the following iteration gives better results with respect to roundoff error:

Choose $x^{(0)}$, an initial vector with n components. Then find $r^{(0)} = b - Ax^{(0)}$ and $p^{(0)} = A'r^{(0)}$. For $i = 0, \ldots, n-1$, compute

$$\alpha_i = \frac{\|A'r^{(i)}\|^2}{\|Ap^{(i)}\|^2} \quad \text{or} \quad \alpha_i = \frac{(Ap^{(i)}, r^{(i)})}{\|(Ap^{(i)}\|^2}$$

$$x^{(i+1)} = x^{(i)} + \alpha_i p^{(i)}$$

$$r^{(i+1)} = r^{(i)} - \alpha_i A p^{(i)}$$

$$\beta_i = \frac{\|A'r^{(i+1)}\|^2}{\|A'r^{(i)}\|^2} \quad \text{or} \quad \beta_i = -\frac{(Ap^{(i)}, AA'r^{(i+1)})}{\|Ap^{(i)}\|^2}$$

$$p^{(i+1)} = A'r^{(i+1)} + \beta_i p^{(i)}.$$

These formulas are derived by using $A'A$ in place of A and $A'r$ in place of r. About twice as much work is required as in the symmetric case.

If A is singular, then $A'A$ is nonnegative definite, $(Ap, p) \geq 0$ for all p, and in this case a least-squares solution can be found. If for any m, $r^{(m)} = 0$ is obtained, then $x^{(m)} = h$ is the solution and the procedure terminates. If $(p^{(m)}, Ap^{(m)}) = 0$, the process also terminates, and a least-squares solution $\tilde{x}^{(m)}$, that is, one which minimizes the residual $b - A\tilde{x}^{(m)}$, can be written as a linear combination of $x^{(0)}, x^{(1)}, \ldots, x^{(m)}$. Thus

$$\tilde{x}^{(m)} = \left[\frac{x^{(0)}}{\|r^{(0)}\|^2} + \frac{x^{(1)}}{\|r^{(1)}\|^2} + \cdots + \frac{x^{(m)}}{\|r^{(m)}\|^2}\right]\left[\frac{1}{\|r^{(0)}\|^2} + \cdots + \frac{1}{\|r^{(m)}\|^2}\right]^{-1}.$$

3.2.4 Stationary Process with Polynomial Iteration Function: The Newton–Raphson Analogue [8, 22, 38]

The simple nonlinear Newton–Raphson-type iteration, $x^{(k+1)} = x^{(k)}(2 - ax^{(k)})$, for finding the reciprocal $1/a$ may also be applied to matrices:

Let

$$X^{(k+1)} = X^{(k)}(2I - AX^{(k)})$$

with $X^{(0)}$ some matrix approximating A^{-1}. (If no better approximation is available, $X^{(0)} = I$ may be tried.) The process converges if and only if the eigenvalues of $(I - AX^{(0)})$ are all of absolute value less than 1. Convergence, when it occurs, is second-order. The iteration could be terminated whenever the sum of the absolute values of the off-diagonal elements of $AX^{(k)}$ is $<\epsilon$. The rate of convergence depends, of course, upon the choice of $X^{(0)}$.

Defects of the method are (a) the difficulty in finding an initial $X^{(0)}$ for which the iteration will converge, and (b) the two matrix multiplications or $2n^3$ multiplications required for each iteration.

3.3 LINEAR ITERATIONS

3.3.1 Point Iterative (Single-Step, Explicit) Methods

The methods obtained by letting the directions $p^{(k)}$ be those of the coordinate axes ϵ_i(where ϵ_i has unity as its ith component and zero as its other $n - 1$ components) are called the single-step methods. Such methods are equivalent to solving each of the original equations in turn for a single unknown, hence the solution vector is changed one component at a time. The sequence in which the coordinate-axis directions are used must be determined. Single-step methods may be written generally in the form

$$x^{(k+1)} = x^{(k)} + \alpha^{(k)}\epsilon_{i(k)}.$$

3.3.1.1 Stationary

Jacobi method: iteration by simultaneous displacements **[7, 11, 13, 27, 63, 67].** The Jacobi method may be written in several forms. First write

$$x^{(nk+i+1)} = x^{(nk+i)} + \alpha^{(nk+i)}\epsilon_i \quad \text{with} \quad \alpha^{(nk+i)} = \frac{(r^{nk}, \epsilon_i)}{(\epsilon_i, A\epsilon_i)} = \frac{r_i^{nk}}{a_{ii}},$$

or

$$x^{(nk+i+1)} = x^{(nk+i)} + \left(\frac{r_i^{(nk)}}{a_{ii}}\right) \cdot \epsilon_i.$$

The Jacobi iteration may also be expressed as

$$x^{(k+1)} = x^{(k)} + C^{(k)}r^{(k)}$$

with $C^{(k)} = C$ and $C = \text{diagonal}\ (1/a_{11}, 1/a_{22}, \ldots, 1/a_{nn})$, where it is assumed that $a_{ii} \neq 0$ for any $i = 1, \ldots, n$ and where $C^{(k)}$ is now independent of k. It can be seen that the process is both stationary and linear. The method consists of solving the ith equation of the system $Ax = b$ for $x_i^{(k+1)}$ using the values $x_j^{(k)}$, $i \neq j$ for the remaining variables, that is

$$x_i^{(k+1)} = \frac{1}{a_{ii}}\left(b_i - \sum_{\substack{j=1 \\ j\neq i}}^{n} a_{ij}x_j^{(k)}\right) \quad \text{for} \quad i = 1, \ldots, n.$$

Finally, write $A = D + L + U$, where D is the diagonal matrix whose diagonal elements are those of A, L is the lower triangular portion of A minus the diagonal, and U is the upper triangular portion of A minus the diagonal. The Jacobi iteration for the system $Ax = b$ then takes the form

$$Dx^{(k+1)} + (L + U)x^{(k)} = b \quad \text{or} \quad x^{(k+1)} = -D^{-1}(L + U)x^{(k)} + D^{-1}b,$$

which converges if and only if $[-D^{-1}(L + U)]^k \to 0$ as $k \to \infty$, that is, if and only if the eigenvalues of $[-D^{-1}(L + U)]$ are less than 1 in absolute value. The matrix $-D^{-1}(L + U) = (I - D^{-1}A)$ is the iteration matrix of the Jacobi process.

The process must be started, of course, with some initial estimate $x^{(0)}$. In the absence of a better estimate for $x^{(0)}$ the approximation $x_i^{(0)} \approx b_i/a_{ii}$ could be used.

Since the error vector $e^{(k+1)}$ at the $(k + 1)$th iteration may be written as before in the familiar form

$$e^{(k+1)} = (I - CA)^{k+1}e^{(0)},$$

with iteration matrix

$$(I - CA) = (I - D^{-1}A),$$

the Jacobi iteration converges for every $x^{(0)}$ and for any order of the equations to the solution of $Ax = b$:

1. If and only if $(I - CA)^{k+1} \to 0$ as $k \to \infty$.
2. If and only if every eigenvalue of $(I - CA)$ is less than 1 in absolute value (see Theorem 30, Appendix B). If 1 is an eigenvalue, that is, $|A| = 0$, then the iteration $x^{(k+1)} = (I - CA)x^{(k)} = (I - CA)^{k+1}x^{(0)}$ converges for every starting point to a solution of the homogeneous system $Ax = 0$, provided all other eigenvalues of $(I - CA)$ are less than 1 in absolute value and the rank of A is $(n - s)$, where s is the multiplicity of the eigenvalue 1.

3. If

$$\sum_{k=1}^{n} \sum_{\substack{i=1 \\ i \neq k}}^{n} \left(\frac{a_{ik}}{a_{ii}}\right)^2 < 1$$

(Theorem of E. Schmidt–Mises–Geiringer).

4. If

$$\sum_{\substack{m=1 \\ k \neq m}}^{n} \left|\frac{a_{km}}{a_{kk}}\right| < 1 \quad \text{for all} \quad k = 1, \ldots, n$$

(Theorem of Frobenius–Mises–Geiringer)
or if

$$\sum_{\substack{k=1 \\ k \neq m}}^{n} \left|\frac{a_{km}}{a_{kk}}\right| < 1 \quad \text{for all} \quad m = 1, \ldots, n$$

(Theorem of Frobenius–Mises–Geiringer).

5. If A is irreducible and
with

$$\sum_{\substack{k=1 \\ i \neq k}}^{n} \left|\frac{a_{ik}}{a_{ii}}\right| \leq 1 \quad \text{for all} \quad i = 1, \ldots, n$$

$$\sum_{\substack{k=1 \\ i \neq k}}^{n} \left|\frac{a_{ik}}{a_{ii}}\right| < 1$$

for at least one value of i.

6. If A is irreducible and

$$\sum_{\substack{i=1 \\ i \neq k}}^{n} \left|\frac{a_{ik}}{a_{ii}}\right| \leq 1 \quad \text{for all} \quad k = 1, \ldots, n$$

with

$$\sum_{\substack{i=1 \\ i \neq k}}^{n} \left|\frac{a_{ik}}{a_{ii}}\right| < 1$$

for at least one value of k.

7. If

$$\left(\frac{l+1}{u+1}\right)^n \gtrless \left(\frac{1}{u}\right) \quad \text{according as } l \gtrless u,$$

where

$$l = \max_{i>j} \left|\frac{a_{ij}}{a_{ii}}\right|, \qquad u = \max_{i<j} \left|\frac{a_{ij}}{a_{jj}}\right|$$

(Theorem of Stein–Rosenberg).

COLLATZ' ESTIMATION OF ERROR If

$$\sum_{\substack{k=1\\ i\neq k}}^{n} |a_{ik}| < |a_{ii}| \quad \text{for all} \quad i,$$

then the following relation holds between the approximation vectors $x^{(k)}$ and the components $e_i^{(k+1)}$ of their errors:

$$|e_i^{(k+1)}| = |x_i^{(k+1)} - x_i| \leq \left(\frac{m}{1-m}\right) \max_i |x_i^{(k+1)} - x_i^{(k)}|,$$

where

$$m = \max_i \frac{\sum_{\substack{k=1\\ i\neq k}}^{n} |a_{ik}|}{|a_{ii}|} < 1.$$

Seidel iteration by successive displacement **[7, 11, 13, 18, 67, 84, 102].** The Seidel iteration may be written in any one of the following ways:

1.

$$x^{(nk+i+1)} = x^{(nk+i)} + \alpha^{(nk+i)}\epsilon_i, \quad \text{with} \quad \alpha^{(nk+i)} = \frac{(r^{nk+1}, \epsilon_i)}{(\epsilon_i, A\epsilon_i)} = \frac{r_i^{(nk+1)}}{a_{ii}},$$

or

$$x^{(nk+i+1)} = x^{(nk+i)} + \left(\frac{r_i^{(nk+1)}}{a_{ii}}\right)\epsilon_i.$$

2.

$$x_i^{(k+1)} = \frac{1}{a_{ii}}\left[b_i - \sum_{j=i+1}^{n} a_{ij}x_j^{(k)} - \sum_{j=1}^{i-1} a_{ij}x_j^{(k+1)}\right].$$

3. Write $A = D + L + U$, where D is the diagonal matrix composed of the diagonal elements of A, L is the lower triangular portion of A minus the diagonal, and U is the upper triangular portion of A minus the diagonal. Then the Seidel iteration for solving $Ax = b$ is described by

$$(D + L)x^{(k+1)} + Ux^{(k)} = b,$$

or

$$x^{(k+1)} = -(D + L)^{-1}Ux^{(k)} + (D + L)^{-1}b.$$

4. In the general iteration and error formulas, $x^{(k+1)} = x^{(k)} + Cr^{(k)}$ and $e^{(k+1)} = (I - CA)^{k+1}e^{(0)}$, with iteration matrix $(I - CA)$, it is clear from (3) that for the Seidel iteration $C = (D + L)^{-1}$ since

$$A = (D + L) + U, \qquad (D + L)^{-1}A = I + (D + L)^{-1}U,$$

and

$$-(D + L)^{-1}U = (I - (D + L)^{-1}A).$$

Hence the iteration matrix may be written

$$-(D + L)^{-1}U = (I - (D + L)^{-1}A).$$

The Seidel iteration which is linear and stationary, assumes values for $x_2^{(0)}, \ldots, x_n^{(0)}$ and computes $x_1^{(1)}$ from the first equation, uses $x_1^{(1)}, x_3^{(0)}, \ldots, x_n^{(0)}$ in the second equation to compute $x_2^{(1)}$, etc., cycling through the equations. In general $x_i^{(k+1)}$ is computed from the ith equation using $x_1^{(k+1)}, \ldots, x_{i-1}^{(k+1)}, x_{i+1}^{(k)}, \ldots, x_n^{(k)}$. It is not necessary that the equations be used in the order $1, 2, \ldots, n$. However, the order must remain fixed in order for the iteration matrix to remain constant, hence for the general results on convergence for constant iteration matrices to apply. The equations must be ordered so that $a_{ii} \neq 0$ for any i.

The Seidel iteration converges for every $x^{(0)}$, and for any order of the equations, to the solution of $Ax = b$ whenever any of the convergence conditions 1, 2, 4, 5, and 6, as stated in Section 3.3.1.1 hold. Collatz' estimation of error [see Section 3.3.1.1] also applies here. Further convergence criteria is contained in Reich's theorem.

REICH'S THEOREM: If A is real symmetric (or Hermitian) and $a_{ii} > 0$ for all i, then the Seidel iteration converges for every starting vector if and only if A is positive definite. [Note that if the system $Ax = b$ is replaced by the system $A'Ax = A'b$ or by $(A^*)'Ax = (A^*)'b$, the conditions of the theorem are satisfied. The procedure is not recommended, however, due to the extra matrix multiplication involved.]

The approximation $x_i^{(0)} = b_i/a_{ii}$ may be used as an initial vector when no better estimate is available.

Drawbacks of the Seidel process are that convergence is only linear, (though it converges about twice as fast as the Jacobi iteration) and that when eigenvalues of the iteration matrix are complex, convergence is irregular. In the case where eigenvalues of the iteration matrix are all real and distinct in absolute value, then an acceleration procedure, such as given in Section 3.4, can be applied.

Note that the Seidel method is a special case of successive overrelaxation (see Section 3.3.1.2) with fixed $\omega = 1$.

Back-and-forth Seidel **[13, 27, 60].**

The back-and-forth Seidel iteration was designed by Aitken and Rosser to overcome one of the objections to the Seidel method. If the iteration matrix for the Seidel method has a compex-conjugate pair of dominant eigenvalues (or even $2q \leq n$ dominant eigenvalues), convergence will be irregular. The iteration matrix for the back-and-forth Seidel method has real eigenvalues.

Assume A is symmetric and definite. Start with the first approximation vector $x^{(0)}$ and find $x^{(1)}$ by the regular Seidel process. Then find $x^{(2)}$ by applying the Seidel process to the equations in reverse order. Find $x^{(3)}$ by the regular Seidel process, etc. To apply the reverse Seidel, let $j = n, n-1, \ldots, 1$ and solve the jth equation for $x_j^{(k+1)}$ using $x_p^{(k+1)}$, $p > j$ obtained from the previous $(n-j)$ equations and the $x_p^{(k)}$, $p < j$ for the remaining variables. The back-and-forth Seidel may be written

$$x^{(nk+i+1)} = x^{(nk+i)} + \alpha^{(nk+i)}(\epsilon_i) \qquad \text{for } k \text{ even}$$

$$x^{(nk+i+1)} = x^{(nk+i)} + \alpha^{(nk+i)}(\epsilon_{n-i}) \qquad \text{for } k \text{ odd}$$

with

$$\alpha^{(nk+i)} = \frac{(r^{nk+1}, \epsilon_i)}{(\epsilon_i, A\epsilon_i)} = \frac{r_i^{(nk+1)}}{a_{ii}} \quad \text{for } k \text{ even}$$

and

$$\alpha^{(nk+i)} = \frac{(r^{nk+1}, \epsilon_{n-i})}{(\epsilon_{n-i}, A\epsilon_{n-i})} = \frac{r_{n-i}^{(nk+1)}}{a_{n-i,n-i}} \quad \text{for } k \text{ odd.}$$

If $A = D + L + U$, where D is the diagonal of A, L is the lower triangular portion of A below the diagonal, and U is the portion of A above the diagonal, then

$$(D + L)\tilde{x}^{(k)} + Ux^{(k)} = b$$

is solved for $\tilde{x}^{(k)}$, and

$$(D + U)x^{(k+1)} + L\tilde{x}^{(k)} = b$$

is solved for $x^{(k+1)}$ after substituting $\tilde{x}^{(k)}$ from the first equation into the second. Thus

$$(D + U)x^{(k+1)} = -L(D + L)^{-1}(b - Ux^{(k)}) + b,$$

and

$$x^{(k+1)} = (D + U)^{-1}\{L(D + L)^{-1}Ux^{(k)} + (I - L(D + L)^{-1})b\}.$$

The iteration matrix is then

$$(D + U)^{-1}L(D + L)^{-1}U.$$

Since A is assumed symmetric, $L = U'$, $L' = U$ and the iteration matrix is

$$(D + U)^{-1}U'(D + U')^{-1}U = (D + U)^{-1}U'[(D + U)^{-1}]'U.$$

For A symmetric and positive definite, Fox [13] pp. 195–197, has proved that the method converges and that the iteration matrix has real eigenvalues λ in the range $0 \leq \lambda < 1$.

de la Garza's method. Let

$$x^{(k+1)} = x^{(k)} + \alpha^{(k)}\epsilon_{i(k)},$$

where $\epsilon_{i(k)} = \epsilon_i$, $k \equiv i \bmod n$ or

$$x^{(nk+i+1)} = x^{(nk+i)} + \alpha^{(nk+i)}\epsilon_i.$$

Using the optimum α found for the measure $\phi_2(x) = (r, r)$ in Section 2.1.1.2 gives

$$\alpha^{nk+i} = \frac{(A\epsilon_i, r^{nk})}{\|A\epsilon_i\|^2},$$

which defines de la Garza's iteration.

Kacmarz' method [3]. Let

$$x^{(k+1)} = x^{(k)} + \alpha^{(k)}A'\epsilon_i,$$

where $k \equiv i \bmod n$, and $p^{(k)} = A'\epsilon_i$. Using the optimum α for measure $\phi_1(x) = (r, (AA')^{-1}r)$ from Section 2.1.1.2 provides the result

$$\alpha^{(k)} = \frac{(\epsilon_i, r^{(k)})}{\|A'\epsilon_i\|^2} = \frac{r_i^{(k)}}{\|A'\epsilon_i\|^2}.$$

Each equation

$$\sum_{j=1}^{n} a_{ij}x_j = b_i$$

of the system $Ax = b$ represents a hyperplane H_i in n space. From a point $x^{(k)}$, representing an approximate solution of $Ax = b$, move to a point $x^{(k+1)}$ which is the foot of the perpendicular from $x^{(k)}$ to one of the hyperplanes H_i. The equation

$$x_j^{(k+1)} = x_j^{(k)} + \alpha^{(k)}a_{ij}$$

states that the segment from $x^{(k)}$ to $x^{(k+1)}$ has direction perpendicular to every line in H_i; the equation

$$\sum_{j=1}^{n} a_{ij}\, x_j^{(k+1)} = b_i$$

states that $x^{(k+1)}$ lies on H_i. Since $x_j^{(k)}$ is known, the two equations may be solved for $\alpha^{(k)}$ to obtain

$$\alpha^{(k)} = \frac{b_i - \sum_{j=1}^{n} a_{ij}x_j^{(k)}}{\sum_{j=1}^{n} a_{ij}^{\,2}} = \frac{r_i^{(k)}}{\|A'\epsilon_i\|^2}$$

(Note $A'\epsilon_i \equiv$ vector forming the ith row of A.)

The method of Kacmarz then is to choose a point $x^{(k)}$, project it on the first hyperplane, project the resulting point onto the second hyperplane, etc., in cyclic order. Each image point must be closer to the solution than the point projected, for the segment from the solution to the image point is a leg of a right triangle whose hypotenuse is a segment from the solution to the point projected.

3.3.1.2 May Be Either Stationary or Nonstationary

Relaxation **[8, 12, 19].** The method of relaxation always takes $p^{(k)}$ to be one of the coordinate axes, ϵ_i, but the particular ϵ_i chosen is decided at the time of computation according to some criteria. One method is to choose ϵ_i when the largest component of the residual vector $r^{(k)}$ is $r_i^{(k)}$, another is to relax that component i for which $(r_i^{(k)})^2/a_{ii}$ is maximal, thereby causing maximal decrease in the error measure ϕ_3 [Sections 3.1 and 3.2.1.1(b)]. Thus

$$x^{(k+1)} = x^{(k)} + \alpha^{(k)}\epsilon_i$$

where

$$\alpha^{(k)} = \frac{(r^{(k)}, \epsilon_i)}{(\epsilon_i, A\epsilon_i)} = \frac{r_i^{(k)}}{a_{ii}}.$$

Relaxation is considered a "hand" method since the choice of ϵ_i must often be made by an experienced operator; the method is usually nonstationary. Relaxation is actually a Seidel process in which the equations are solved in varied (rather than fixed) order, so that faster convergence is achieved.

If the coefficient matrix A is symmetric and positive definite, the relaxation method will converge to the solution for any starting vector.

Extrapolated Jacobi **[13].** The Jacobi iteration was representable as

$$Dx^{(k+1)} + (L + U)x^{(k)} = b$$

$$x^{(k+1)} = -D^{-1}(L + U)x^{(k)} + D^{-1}b$$

where $A = D + L + U$ and $-D^{-1}(L + U) = I - D^{-1}A$. Thus $x^{(k+1)} = (I - D^{-1}A)x^{(k)} + D^{-1}b$ represents the original Jacobi method. Now introduce a parameter ω and consider the iteration defined by

$$x^{(k+1)} = (I - \omega D^{-1}A)x^{(k)} + \omega D^{-1}b$$

and termed an extrapolated Jacobi iteration. It should be possible to find ω so that the largest absolute eigenvalue μ of $(I - \omega D^{-1}A)$ is smaller than the largest absolute eigenvalue $\tilde{\mu}$ of $(I - D^{-1}A)$. Since $D^{-1}A$ has a unit diagonal, it follows that

$$(1 - \mu) = \omega(1 - \tilde{\mu})$$

or

$$\mu = 1 - \omega + \omega\tilde{\mu}.$$

Suppose A is symmetric, positive definite so that $B = D^{-1}A$ is also. Let λ be an eigenvalue of B. Then $\tilde{\mu} = 1 - \lambda$ and $\mu = 1 - \omega\lambda$ and we must have $|1 - \lambda| < 1$ for convergence of Jacobi, $|1 - \omega\lambda| < 1$ for convergence of extrapolated Jacobi. Hence the Jacobi converges for $\lambda_{\max} < 2$ and the extrapolated Jacobi converges for $\omega\lambda_{\max} < 2$. The best choice for ω is that for which $1 - \lambda_{\min}\,\omega = -(1 - \lambda_{\max}\omega)$ so that then the largest values of μ have equal and opposite signs. Then

$$\omega = \frac{2}{\lambda_{\min} + \lambda_{\max}} \quad \text{and} \quad \mu = |1 - \lambda\omega|$$

$$= \left|1 - \frac{2\lambda}{\lambda_{\min} + \lambda_{\max}}\right| \le \left|\frac{\lambda_{\max} - \lambda_{\min}}{\lambda_{\max} + \lambda_{\min}}\right| = \left|\frac{\dfrac{\lambda_{\max}}{\lambda_{\min}} - 1}{\dfrac{\lambda_{\max}}{\lambda_{\min}} + 1}\right| < 1.$$

The extrapolation is most effective when $\lambda_{\max}/\lambda_{\min} \sim 1$, since in this case μ is small.

The possibility of using different ω at successive steps has been considered, yielding a nonstationary iteration. In particular see Richardson's method in Section 3.3.1.3. Note that the extrapolated Jacobi is a linear (Chebyshev) acceleration of the Jacobi method with fixed ω (see Section 3.4.4.1).

Extrapolated Seidel **[13].** The Seidel iteration was represented as

$$(D + L)x^{(k+1)} + Ux^{(k)} = b \quad \text{where} \quad A = D + L + U,$$

which may be written

$$D(x^{(k+1)} - x^{(k)}) = b - Ux^{(k)} - Dx^{(k)} - Lx^{(k+1)}$$

or

$$D(x^{(k+1)} - x^{(k)}) = b - Ax^{(k)} - L(x^{(k+1)} - x^{(k)}).$$

Introduce the parameter ω^{-1} and define the iteration

$$\omega^{-1}D(x^{(k+1)} - x^{(k)}) = b - Ux^{(k)} - Dx^{(k)} - Lx^{(k+1)},$$

$$(\omega^{-1}D + L)x^{(k+1)} = b - Ux^{(k)} - (1 - \omega^{-1})Dx^{(k)},$$

$$x^{(k+1)} = [I - (\omega^{-1}D + L)^{-1}A]x^{(k)} + (\omega^{-1}D + L)^{-1}b.$$

The iteration matrix is $[I - (\omega^{-1}D + L)^{-1}A]$ and we are interested in determining ω so as to give this matrix a small maximum eigenvalue.

The method is identical to the method of successive overrelaxation described in the following section, where the optimum value of ω is found for matrices satisfying property A and discussed for other cases. The possibility of using different ω at successive steps has also been considered, giving a nonstationary iteration.

***Successive overrelaxation* (*SOR*) [12, 40, 50, 111, 113]** Assume that A is symmetric and positive definite and consider the iteration

$$x^{(k+1)} = x^{(k)} + \alpha^{(k)}\epsilon_i.$$

Suppose that $\alpha^{(k)}$ is taken to be ω_k times the $\alpha^{(k)}$ for which the error measure $\phi(x^{(k)} + \alpha^{(k)}\epsilon_i)$ is minimized. Then, using the measure $\phi_3(x) = (r, Sr) = (r, (A')^{-1}r)$, it has been shown [see Section 3.2.1.1(b)] that $\phi_3(x^{(k)} + \alpha^{(k)}\epsilon_i)$ is minimized when $\alpha^{(k)} = r_i^{(k)}/a_{ii}$. Thus the difference in error measures produced by $x^{(k)}$ and $x^{(k)} + \alpha^{(k)}\epsilon_i$ is

$$\begin{aligned}\phi_3(x^{(k)}) - \phi_3(x^{(k)} + \alpha^{(k)}\epsilon_i) &= 2\alpha^{(k)}(A^{-1}r^{(k)}, A\epsilon_i) - (\alpha^{(k)})^2(\epsilon_i, A\epsilon_i)\\ &= 2\alpha^{(k)}r_i^{(k)} - (\alpha^{(k)})^2 a_{ii}\\ &= \frac{\omega_k(2-\omega_k)(r_i^{(k)})^2}{a_{ii}}\end{aligned}$$

when $\alpha^{(k)}$ is ω_k times the optimal $\alpha^{(k)}$, that is, $\alpha^{(k)} = \omega_k r_i^{(k)}/a_{ii}$, and the error is decreased provided $0 < \omega_k < 2$. The parameter ω_k is called a *relaxation* parameter. The iteration defined is linear and single-step and is stationary when $\omega_k = \omega$ for all k, otherwise it is nonstationary.

For $\omega = 1$ the method is the Seidel iteration [see Section 3.3.1.1(b)].
For $0 < \omega < 1$ the method is termed *underrelaxation.*
For $1 < \omega < 2$ the method is termed *overrelaxation.*

Successive overrelaxation (SOR) was developed by Young and Frankel (1950) and has proved particularly useful in solving systems of linear equations arising from difference equations for solution of elliptic partial differential equations. The resulting matrices usually have special properties which promote the effectiveness of iterative methods of solution. One such property is defined below as well as in the Glossary.

PROPERTY A A matrix $A = (a_{ij})$ has property A if there exist two disjoint subsets S and T of the first n positive integers whose sum is the first n positive integers such that for any $a_{ij} \neq 0$ either

1. $i = j$, or
2. i is in S and j is in T, or
3. i is in T and j is in S.

This is equivalent to stating that by appropriate permutations of its rows and corresponding columns, the matrix A can be written in the form

$$A = \begin{pmatrix} D_1 & F \\ G & D_2 \end{pmatrix},$$

where the D_i are square diagonal matrices and F and G are arbitrary rectangular matrices.

Stationary successive overrelaxation for solving $Ax = b$, (i.e., all $\omega_k = \omega$), may be written as follows:

$$x_i^{(k+1)} = -\left(\frac{\omega}{a_{ii}}\right)\left(\sum_{j=1}^{i-1} a_{ij}x_j^{(k+1)} + \sum_{j=i+1}^{n} a_{ij}x_j^{(k)} - b_i\right) - (\omega - 1)x_i^{(k)}$$

for $i = 1, \ldots, n$, or, in matrix notation,

$$x^{(k+1)} = -\omega D^{-1}(Lx^{(k+1)} + Ux^{(k)} - b) - (\omega - 1)Ix^{(k)},$$

$$x^{(k+1)} = [I - (\omega^{-1}D + L)^{-1}A]x^{(k)} + (\omega^{-1}D + L)^{-1}b,$$

where

$$A = D + L + U,$$

$$D = (a_{ii}), \qquad L = (a_{ij}), \qquad i > j, \qquad U = (a_{ij}), \qquad i < j,$$

for $i, j = 1, \ldots, n$. The iteration matrix for the stationary SOR method is then

$$T = [I - (\omega^{-1}D + L)^{-1}A].$$

The following theorems are available on convergence of the SOR method, optimal value of ω, and the relation of the SOR process to the Jacobi and Seidel iterations:

THEOREM 1: Let A be symmetric, positive definite, and irreducible. Then SOR converges if and only if $0 < \omega < 2$.

THEOREM 2: Let $a_{ij} = a_{ji} \leq 0$, $a_{ii} > 0$ and $0 < \omega < 2$. Then SOR converges if and only if A is positive definite.

THEOREM 3: Let A be symmetric, positive definite, irreducible, and satisfy property A. Let $\tilde{\lambda}$ denote an eigenvalue of the Jacobi iteration matrix and λ its largest absolute eigenvalue. Let $\tilde{\mu}_\omega$ denote an eigenvalue of the SOR iteration matrix T_ω, and $\mu_\omega = \lambda(T_\omega)$ its largest absolute eigenvalue. Then the relationship between $\tilde{\lambda}$ and the corresponding $\tilde{\mu}_\omega$ is

$$(\tilde{\mu}_\omega + \omega - 1) = \omega\tilde{\lambda}\sqrt{\tilde{\mu}_\omega}.$$

the optimal value of ω is

$$\omega_{\text{opt}} = 1 + \left(\frac{\lambda}{1 + \sqrt{1 - \lambda^2}}\right)^2 = \frac{2}{1 + \sqrt{1 - \lambda^2}},$$

and when $\omega = \omega_{\text{opt}}$, we have $\mu_{\omega_{\text{opt}}} = \lambda(T_{\omega_{\text{opt}}}) = \omega_{\text{opt}} - 1$. The rate of convergence of SOR with $\omega = \omega_{\text{opt}}$ is about $2\sqrt{2}$ times the square root of the rate of convergence of the Jacobi method and about twice the square root of the rate of convergence of the Seidel method. In order to compute ω_{opt} in the case where property A is satisfied it is first necessary to estimate λ, the largest absolute eigenvalue of the Jacobi iteration matrix. Now when $\omega = 1$, the Seidel iteration is obtained, so that letting $\omega = 1$ in the above relationship between $\tilde{\lambda}$ and $\tilde{\mu} = \tilde{\mu}_1$ gives $\mu_1 = \lambda\sqrt{\mu_1}$, or $\mu_1 = \lambda^2$. Hence performing the SOR iteration for $\omega = 1$ and using the successive solutions $x^{(k-1)}$, $x^{(k)}$ to compute the differences $y^{(k-1)} = x^{(k)} - x^{(k-1)}$ gives a possible means of estimating $\mu_1 = \lambda^2$, since it may be shown that

$$\frac{\|y^{(k)}\|}{\|y^{(k-1)}\|} \to \lambda^2 \quad \text{as} \quad k \to \infty.$$

Any norm may be used in this computation. A decrease in the convergence rate corresponding to an overestimate of ω_{opt} is not too serious; however, an underestimate of ω_{opt} causes a much greater relative decrease in the convergence rate.

When A has property A, the value of ω_{opt} is independent of the order in which the unknowns are determined for a great many orderings.

THEOREM 4: Let $a_{ij} = a_{ji} \leq 0$ and $a_{ii} > 0$. If SOR converges for any one ω, $0 < \omega < 2$, then the Jacobi process converges. If the Jacobi iteration converges then SOR converges for all ω such that $0 < \omega < 2$. When A is also symmetric and positive definite, then there exists an ω such that

$$\sqrt{R(T_1)} \leq R(T_\omega) \leq 2\sqrt{R(T_1)},$$

where $R(T_1)$ is the convergence rate of the Seidel method, that is, $\omega = 1$.

When property A is *not* satisfied, the optimal value of ω is not known and is not always the same as the ω_{opt} defined above in Theorem 3. It must usually be found experimentally. It has been shown by Kahan that for the optimal value of ω, denoted here simply as ω, we have

$$1 < \omega < 2,$$

and

$$\omega_{\text{opt}} - 1 \leq |\lambda(T_\omega)| < \omega_{\text{opt}} - 1.$$

Nonstationary SOR variants may be obtained by varying the parameters ω_k. Improved average rates of convergence may be obtained by use of the cyclic Chebyshev acceleration described in Section 3.4.4. The resulting iterations are often called Chebyshev semi-iterative methods; they improve the average rate of convergence, but not the asymptotic rate.

3.3.1.3 Nonstationary

***Richardson's method* [12, 13, 58, 60, 110].** Assume that the system $Ax = b$ is to be solved, for A symmetric and positive definite, by means of the iteration

$$x^{(k+1)} = x^{(k)} + \alpha_k r^{(k)},\ x^{(0)} \quad \text{arbitrary},$$

where $r^{(k)} = b - Ax^{(k)}$ is the kth residual. The kth error vector is given by

$$\begin{aligned} e^{(k)} &= h - x^{(k)} = A^{-1}b - x^{(k-1)} - \alpha_{k-1}(b - Ax^{(k-1)}) \\ &= e^{(k-1)} - \alpha_{k-1}A(h - x^{(k-1)}) = e^{(k-1)}(I - \alpha_{k-1}A). \end{aligned}$$

It will be assumed now that α_{k-1} is *dependent* on k, that is, that the process is nonstationary. Then

$$e^{(k)} = e^{(k-1)}(I - \alpha_{k-1}A)$$

$$e^{(k)} = e^{(0)} \prod_{j=0}^{k-1} (I - \alpha_j A) = e^{(0)} \cdot P_k(A)$$

where

$$P_k(A) = \prod_{j=0}^{k-1} (I - \alpha_j A)$$

is a polynomial of degree k in A such that $P_k(0) = 1$ and $P_k(1/\alpha_j) = 0$. It is desired that $e^{(k)}$ be made as small as possible, hence that $P_k(A)$ be made as small as possible.

The problem may be formulated in the following manner:

Find a kth-degree polynomial $P_k(x)$ defined for $a' \leq x \leq b'$, so that $P_k(0) = 1$ and such that

$$\max_{a' \leq x \leq b'} |P_k(x)|$$

is a minimum. Such a polynomial was given by W. Markoff (in 1892) and is defined by

$$P_k(x) = \frac{T_k\left[\dfrac{b' + a' - 2x}{b' - a'}\right]}{T_k(y_0)}, \quad \text{where} \quad y_0 = \frac{b' + a'}{b' - a'} > 1,$$

and $T_k(y) = \cos(k \arccos y)$ is the Chebyshev polynomial of degree k adjusted to the interval $-1 \leq y \leq 1$. Then $P_k(x)$ is a Chebyshev polynomial of degree k adjusted to the interval $a' \leq x \leq b'$ and scaled so that $P_k(0) = 1$.

Let

$$y = \frac{b' + a' - 2x}{b' - a'}.$$

Then y_0 is the value of y for $x = 0$, $y = 1$ for $x = a'$, and $y = -1$ for $x = b'$.

$$|P_k(y)| \text{ has maximum value } 1/T_k(y_0),$$

where

$$P_k(x) = \frac{T_k(y)}{T_k(y_0)} \quad \text{and} \quad T_k(y) = \cos(k\theta), \qquad \cos\theta = y.$$

The relaxation factors α_j are then the reciprocals of the zeros of the Chebyshev polynomial $P_k(x)$, or equivalently, the reciprocals of the zeros of $T_k(y)$. The zeros of $T_k(y)$ are

$$y_j^{(k)} = \cos\left(\frac{(2j+1)\pi}{2k}\right) \quad \text{for} \quad j = 0, \ldots, k-1.$$

Now

$$y = \frac{b' + a' - 2x}{b' - a'}, \quad \text{so that} \quad x = \frac{(b' + a') - y(b' - a')}{2}.$$

Thus

$$\alpha_j = 2[(b' + a') - y_j^{(k)}(b' - a')]^{-1}, \qquad j = 0, \ldots, k-1,$$

which now defines the iteration $x^{(k+1)} = x^{(k)} + \alpha_k r^{(k)}$ for given a' and b'. It will be shown later that the best values of a' and b' to use would be the minimum and maximum eigenvalues of A, respectively.

In the case $k = 1$ we have $\alpha_0 = 2/(b' + a')$, the value of ω obtained in the extrapolated Jacobi procedure, Section 3.3.1.2(b), which is identical to the Richardson method when $k = 1$ and a' and b' are the minimum and maximum eigenvalues of A, respectively.

The larger the matrix A, the larger value of k is needed. In his original paper Richardson gave only rough suggestions for determining the α_j. It remained for Flanders and Shortley, Lanczos, and Young to obtain the α_j described above, producing rapid convergence.

The kth-iteration matrix is $T^{(k)} = (I - \alpha_k A)$, and the spectral radius of

$$P_k = P_k(A) = \prod_{i=0}^{k-1} T^{(i)}$$

is $\lambda(P_k) = 1/T_k(y_0)$. Hence the average rate of convergence for fixed k is

$$R(P_k) = -\frac{1}{k}\log\frac{1}{T_k(y_0)} = \frac{1}{k}\log T_k(y_0).$$

Note that $T_k(y)$ is valid even when $y > 1$, for although $\cos^{-1} y$ is complex, $T_k(y) = \cosh(k\cosh^{-1} y)$ is real. Since $y_0 > 1$, we have

$$\cosh^{-1} y_0 = \log[y_0 + \sqrt{y_0^2 - 1}],$$

and since

$$\cosh u = \tfrac{1}{2}(e^u + e^{-u}),$$

it follows that

$$T_k(y_0) = \tfrac{1}{2}[(y_0 + \sqrt{y_0^2 - 1})^k + (y_0 + \sqrt{y_0^2 - 1})^{-k}].$$

It will be seen from the remarks below that an accurate calculation of α_j depends on good estimates of the maximum and minimum eigenvalues of A. Let λ_i be the ith eigenvalue of A and let v_i be the corresponding eigenvector. Since A is symmetric, the v_i are linearly independent by Theorem 39, Appendix B. By Theorem 70, Appendix B, the error vector $e^{(0)}$ can be written uniquely in the form

$$e^{(0)} = \sum_{i=1}^{n} c_i v_i,$$

where the c_i are scalars. Then

$$e^{(k)} = \sum_{i=1}^{n} c_i v_i P_k(\lambda_i).$$

Note that $P_k(\lambda_i)$ is an eigenvalue of $P_k(A)$ by Theorem 43, Appendix B. Now let $a' \le \lambda_i \le b'$ and

$$M_k = \max_i |P_k(\lambda_i)|.$$

Then

$$\|e^{(k)}\|^2 \le \|e^{(0)}\|^2 \, M_k^{\,2}.$$

Thus, ideally, a' should be $\lambda_{\min}$ and b' should be $\lambda_{\max}$ of A. It remains to determine bounds for λ_i. Since A is symmetric positive definite, $\lambda_i > 0$. An upper bound is furnished by

$$\lambda_k \le \max_i \left(\sum_{j=1}^{n} |a_{ij}| \right),$$

or see Theorem 55, Appendix B, for other bounds.

With the α_j given above and assuming A is symmetric and positive definite the method converges for any choice of k. Or if $0 < \alpha_j < 2/b'$, for all j, then the method converges, since

$$|1 - \alpha_j \lambda| < 1, \quad \text{where} \quad 0 < \lambda \le b'.$$

To carry out the iteration, first choose some value of k, say $k = 20$, and compute α_j, $j = 1, \ldots, 20$ at the start. Do $k = 20$ iterations using α_1 for the first iteration, α_2 for the second, etc. Compute the error, $e^{(20)}$. If $e^{(20)}$ is not small enough, repeat the cycle using the same values of α_j and with $x^{(20)}$ as the starting value, $x^{(0)}$.

The order in which the α_j are used is arbitrary and does not affect the convergence, but may affect roundoff error. Suppose the α_j are ordered according to magnitude, that is, so that $|\alpha_1| \ge |\alpha_2| \ge |\alpha_3| \geqq \cdots \geqq |\alpha_{k-1}|$. It is found that poor roundoff results are obtained when the α_j are used in descending order of magnitude; better results are obtained when α_j are used in ascending magnitude, and even better results are obtained when the procedure uses alternately large and small values of α_j, for example,

for k even, use $\alpha_{\frac{1}{2}k+1}, \alpha_{\frac{1}{2}k}, \alpha_{\frac{1}{2}k+2}, \alpha_{\frac{1}{2}k-1}, \ldots, \alpha_k, \alpha_1$,

for k odd, use $\alpha_{\frac{1}{2}(k+1)}, \alpha_{\frac{1}{2}(k+1)+1}, \alpha_{\frac{1}{2}(k+1)-1}, \ldots, \alpha_k, \alpha_1$.

***Lanczos' method* [23, 48, 78, 79]**

A IS HERMITIAN POSITIVE DEFINITE Given the system $Ax = b$ where A is Hermitian (or symmetric) positive definite. A then has real and positive eigenvalues, so assume $\lambda_{\max} \leq \mu$. Such a bound may be computed from Theorem 55, Appendix B. Now rewrite $A = A/\mu$, $b = b/\mu$ so that the new system $Ax = b$ has eigenvalues between 0 and 1. Assume that the eigenvectors v_i are also normalized, that is, $Av_i = \lambda_i v_i$ and $(v_i, v_j) = \delta_{ij}$. Now from Theorems 39 and 70 of Appendix B, b may be written as a linear combination of the eigenvectors:

$$b = \sum_{i=1}^{n} \beta_i v_i$$

where $\beta_i = (b, v_i)$. Hence

$$x = \sum_{i=1}^{n} \frac{\beta_i v_i}{\lambda_i}.$$

If $P_m(x)$ is a polynomial of degree m which approximates $1/x$, then

$$x \sim \sum_{i=1}^{n} \beta_i P_m(\lambda_i) v_i = P_m(A)b.$$

The residual is $r^{(m+1)} = b - Ax \sim b - AP_m(A)b = (1 - AP_m(A))b$ which must be made as small as possible. Hence minimize $|1 - xP_m(x)|$ uniformly in the interval $0 < x \leq 1$ so that the difference between $1/x$ and $P_m(x)$ approaches 0. Define $F_{m+1}(x) = 1 - xP_m(x)$ with $F_{m+1}(0) = 1$ and let

$$\epsilon^{(m+1)} = \frac{|r^{(m+1)}|}{|x|}.$$

Now introduce the Chebyshev polynomials normalized to the interval [0, 1],

$$T_n(x) = \cos n\theta$$

with

$$x = \tfrac{1}{2}(1 - \cos\theta) = \sin^2(\tfrac{1}{2}\theta),$$

and set

$$F_{m+1}(x) = \frac{1 - T_{m+2}(x)}{2(m+2)^2 x} = \frac{\sin^2(m+2)(\frac{1}{2}\theta)}{(m+2)^2 \sin^2(\frac{1}{2}\theta)}.$$

For this choice we have $xF_{m+1}(x) \leq (m+2)^{-2}$, hence $\epsilon^{(m+1)} \leqq (m+2)^{-2}$

when $0 \leq x \leq 1$. Then an approximate solution for x is

$$x^{(m)} \sim \left(\frac{1 - F_{m+1}(A)}{A}\right) b = P_m(A)b.$$

Introduce now the polynomials $g_m(x)$ which have integer coefficients:

$$g_m(x) = \frac{(m+2)^2}{4} \frac{(1 - F_{m+1}(x))}{x} = \frac{T_{m+2}(x) + 2(m+2)^2 x - 1}{8x^2}.$$

From the recursion formula for Chebyshev polynomials, the following recursion for $g_m(x)$ is obtained:

$$g_{m+1}(x) = 2(1 - 2x)g_m(x) - g_{m-1}(x) + (m+2)^2,$$

with $g_0(x) = 1$, $g_1(x) = 6 - 4x$. Now we wish to generate the vectors $g_m(A) \cdot b = g_m'$, so define a matrix $B = 2I - 4A$ and write

$$g_{m+1} = Bg_m - g_{m-1} + (m+2)^2 b$$

with $g_0 = b$, $g_1 = Bg_0 + 4b$. Finally the solution is

$$x \sim \frac{4}{(m+2)^2} g_m(A)b = \frac{4g_m}{(m+2)^2} = x^{(m)},$$

and its residual is

$$r^{(m+1)} = b - Ax^{(m)} = (m+2)^{-2} [(m+2)^2 b - 4Ag_m].$$

The method described provides a "purification" technique for cutting down those components of b in the direction of the eigenvectors corresponding to large eigenvalues. Then attention may be focused on the contribution of the smaller eigenvalues. The above process may be continued by using the residual as the new b, or some other method may be used to continue if the solution is not sufficiently accurate. The method is not recommended for matrix inversion because of its dependence on a specific right-hand side, b. Note that the recursion for g_{m+1} in terms of g_m and g_{m-1} can be written as a recursion for $x^{(m+1)}$ in terms of $x^{(m)}$ and $x^{(m-1)}$, so that the iteration is seen to be linear, nonstationary and of *degree* 2. Thus

$$x^{(m+1)} = \frac{(m+2)^2 Bx^{(m)} - (m+1)^2 x^{(m-1)} + 4(m+2)^2 b}{(m+3)^2}.$$

A IS ARBITRARY AND COMPLEX Given the system $Ax = b$ with A an arbitrary complex matrix, let $A_1 = \frac{1}{2}(A + A^{*\prime})$ and $A_2 = -\frac{1}{2}i(A - A^{*\prime})$. Then $A = A_1 + iA_2$, $A^{*\prime} = A_1 - iA_2$, and

$$A^{*\prime}A = (A_1)^2 + (A_2)^2 + i(A_1A_2 - A_2A_1).$$

Let $\lambda_{\max}$ be the largest eigenvalue of $A^{*\prime}A$, λ_1 be the largest eigenvalue of A_1, and λ_2 be the largest eigenvalue of A_2. Then $\lambda_{\max} \leq (\lambda_1 + \lambda_2)^2$. Since λ_1 and λ_2 can be estimated, an upper bound μ for $\lambda_{\max}$ can be found without forming $A^{*\prime}A$.

Now rewrite the system, redefining A and b as $A = A^{*\prime}A/\mu$, $b = A^{*\prime}b/\mu$. Then for the new system, $Ax = b$, A is Hermitian positive definite with real positive eigenvalues between 0 and 1. Following through the formulas for generating g_m it is clear that is never necessary to form $A^{*\prime}A$, since this expression never appears alone. It is always multiplied by a vector, for example, b. Hence perform $A^{*\prime}(Ab)$ in two steps, Ab, then $A^{*\prime}(Ab)$, and so perform two multiplications of a vector by a matrix. The solution is found as before by generating the vectors g_m, giving the approximation $x^{(m)} \sim 4g_m/(m + 2)^2$.

BELLAR'S VARIATION OF THE LANCZOS METHOD A variation of the Lanczos method given by Bellar [48] is purported by its author to improve convergence. Assume the system $Ax = b$ has been transformed as in the last two sections so that A has real positive eigenvalues between 0 and 1, in fact between $x_0 > 0$ and 1. Also assume the eigenvectors v_i are normalized.

Again write

$$b = \sum_{i=1}^{n} \beta_i v_i \quad \text{where} \quad \beta_i = (b, v_i)$$

and

$$x = \sum_{i=1}^{n} \frac{\beta_i v_i}{\lambda_i} \sim P_m(A)b = x^{(m)}.$$

Bellar obtained a polynomial $P_m(x)$ of degree m that approximates $1/x$ on the interval $[x_0, 1]$ in such a way that

$$\max_{0<x_0\leq x\leq 1} |1 - xP_m(x)| = \min_{a_i} \max_{0<x_0\leq x\leq 1} \left| \frac{x^{-1} - \sum_{i=0}^{m} a_i x^i}{x^{-1}} \right| = d_m \cdot *$$

Let

$$\alpha_i = [(1 + \sqrt{x_0})^{2i} + (1 - \sqrt{x_0})^{2i}]$$

$$\cos\theta = \frac{(1 + x_0 - 2x)}{(1 - x_0)}$$

$$d_m = 2(1 - x_0)^{m+1}\alpha_{m+1}^{-1}.$$

Then $1 - xP_m(x) = d_m \cos(m + 1)\theta$, where $P_m(x)$ is the polynomial of degree m that is the solution of *. A simple recursion will be developed for the generation of $P_m(x)$. Write $g_m(x) = \alpha_{m+1}P_m(x)$. It may be shown that

$$g_m(x) = 2(1 + x_0 - 2x)g_{m-1}(x) - (1 - x_0)^2 g_{m-2}(x) + 4\alpha_m$$

with

$$g_1(x) = 16(1 + x_0 - x), \qquad g_0(x) = 4, \qquad g_{-1}(x) = 0.$$

Now define the vectors $g_m \equiv g_m(A)b$, which may be generated recursively by

$$g_{m+1} = [2(1 + x_0)I - 4A]g_m - (1 - x_0)^2 g_{m-1} + 4\alpha_{m+1}b,$$

where $g_{-1} = 0$, $g_0 = 4b$. Then $x^{(m)} = P_m(A)b = \alpha_{m+1}^{-1} g_m(A)b = \alpha_{m+1}^{-1} g_m$ is an approximation to the solution whose residual is

$$r^{(m+1)} = b - Ax^{(m)} = b - \alpha_{m+1}^{-1} A g_m = \frac{1}{\alpha_{m+1}}(\alpha_{m+1} b - A g_m),$$

and whose relative error is

$$\frac{\|x^{(m)} - x\|}{\|x\|} \leq d_m.$$

For $d_m = 0.05$ and $x_0 = 0.001$, the number of iterations required for the Lanczos method was 79, as compared with 59 for the Bellar variation.

Other figures are given in [48].

3.3.1.4 Monte-Carlo Methods [8, 16, 33, 55, 56, 59, 62, 70, 87, 105]

Monte-Carlo methods (first proposed by Von Neumann and Ulam) give statistical estimates for the elements of the inverse of a matrix or for components of the solution vector of a linear system by performing random sampling of a certain chance variable whose expected value (mean value) is the desired solution. Such methods can be useful in any of the following situations:

1. A quick rough estimate of the solution is desired, which will then be refined by other means. The first few steps of a Monte-Carlo process tend to improve results greatly, whereas many additional steps would be needed to refine the result to sufficient accuracy.
2. The problem is too large or too intricate for any other treatment.
3. Just one component of the solution vector or one element of the inverse matrix is desired. In this case it is not necessary to perform as much work as for the complete solution. The requirement might arise in the solution of linear difference equations derived from an elliptic partial differential equation when the solution is wanted at just one mesh point.

The Monte-Carlo method to be described will give a statistical estimate of one component of the solution vector defined as the limit of a stationary linear iteration. Other Monte-Carlo methods are given in [16], and Monte-Carlo methods for an element of the inverse matrix are given in [55, 87].

The general stationary linear iteration for solving the system $Ax = b$ may be written

$$x^{(k+1)} = Tx^{(k)} + q$$

where $q = (I - T)A^{-1}b$. For $T = (I - D^{-1}A)$ and $T = (I - (D + L)^{-1}A)$, where $A = D + L + U$, we obtain the Jacobi and Gauss–Seidel iterations, respectively. The residual corresponding to $x^{(k)}$ is

$$r^{(k)} = q - (I - T)x^{(k)} = x^{(k+1)} - x^{(k)},$$

and it can be easily shown that

$$r^{(k+1)} = Tr^{(k)} = T^{k+1}r^{(0)},$$

and that

$$x^{(k+1)} = x^{(0)} + r^{(0)} + r^{(1)} + \cdots + r^{(k)}$$

$$x^{(k+1)} = x^{(0)} + (I + T + T^2 + \cdots + T^k)r^{(0)}.$$

Define

$$\|T\| = \max_i \sum_{j=1}^{n} |t_{ij}|,$$

and assume that $\|T\| < 1$, so that the iteration converges to its solution, $h = A^{-1}b = (I - T)^{-1}q$. Then

$$h = x^{(0)} + (I - T)^{-1}r^{(0)}$$

and

$$h = x^{(k)} + T^k(I - T)^{-1}r^{(0)}.$$

The following inequalities hold:

$$\|r^{(k+1)}\| \le \|T\|^{k+1}\ \|r^{(0)}\|,$$

$$\|(I - T)^{-1}\| \le (1 - \|T\|)^{-1},$$

$$\|h - x^{(k)}\| \le \frac{\|T\|^k\ \|r^{(0)}\|}{1 - \|T\|}.$$

Now the analogous Monte-Carlo method will give a statistical estimation of one component of $h = x^{(0)} + (I + T + T^2 + \cdots)r^{(0)} = x^{(0)} + (I - T)^{-1}r^{(0)}$ using the same T, $x^{(0)}$ and $r^{(0)}$ as the stationary linear iteration.

(a) Monte-Carlo method for computing one component of the solution vector, $h = x^{(0)} + (I - T)^{-1}r^{(k)}$, ***of*** $Ax = b$. Consider a random walk of the following type: Let $P_1, \ldots, P_n$ be a set of n points. Start from any point P_i and jump from point to point in such a way that the probability of a direct move from P_r to P_s is p_{rs}. At any point P_r there is a probability

$$p_r = 1 - \sum_{s=1}^{n} p_{rs}$$

that the walk ends there. At each transition $P_r \to P_s$ there is an associated transition value v_{rs} defined by $t_{rs} = v_{rs}p_{rs}$, where t_{rs} is the r, s element of the iteration matrix T. Let N be the duration of a random walk, that is the number of steps taken before the walk stops. N is a random variable whose mean and variance, when the walk starts at P_i, have bounds

$$E(N:i_0 = i) \leq \frac{1}{(1 - \|T\|)},$$

$$\sigma^2(N:i_0 = i) \leq \frac{1}{(1 - \|T\|)}.$$

Now define the random variable

$$g = \frac{(v_{i_0 i_1} v_{i_1 i_2} \cdots v_{i_{N-1} i_N}) r_{i_N}^{(0)}}{p_{i_N}}$$

which gives a "score" for a random walk,

$$P_{i_0} \to P_{i_1} \to P_{i_2} \to \cdots \to P_{i_{N-1}} \to P_{i_N},$$

of N steps beginning at P_{i_0} and ending at P_{i_N}. It can be shown that the expected value of the random variable g when $i_0 = i$ is

$$E(g:i_0 = i) = (I - T)^{-1} r_i^{(0)}.$$

Hence if K independent determinations of g for $i_0 = i$ are made and denoted $g_1, g_2, \ldots, g_K$, then their average added to $x_i^{(0)}$ will give an approximation, $\tilde{x}_i^{(K)}$ for h_i. Thus

$$h_i = x_i^{(0)} + E(g:i_0 = i) = x_i^{(0)} + (I - T)^{-1} r_i^{(0)},$$

$$\tilde{x}_i^{(K)} = x_i^{(0)} + \frac{(g_1 + g_2 + \cdots + g_K)}{K}.$$

The variance of g when $i_0 = i$ is bounded by

$$\sigma^2(g:i_0 = i) \leq \frac{\|r^{(0)}\|^2}{2(1 - \|T\|)^2}.$$

To carry out the Monte-Carlo process, the quantities p_{ij} and v_{ij} must first be defined. One good choice (which is also simple) is to let

$$p_{ij} = |t_{ij}|,$$

and

$$v_{ij} = \text{sign } t_{ij},$$

so that

$$p_{ij}v_{ij} = t_{ij},$$

$$p_{ij} \geq 0,$$

and

$$\sum_{j=1}^{n} p_{ij} < 1.$$

Next use a random number generating subroutine to choose a random number ξ between 0 and 1. If P_i is the point to which the walk proceeded, find the smallest integer s, $s = 1, \ldots, n$, for which

$$\sum_{j=1}^{s} p_{ij} > \xi.$$

This sum is the cumulative probability of going from P_i to P_s. If such an s exists, then update the value of g by multiplying it by v_{is}. If

$$\sum_{j=1}^{n} p_{ij} \leq \xi,$$

then the walk stops and the final value of g for the walk is computed by multiplying the current value of g by $r_i^{(0)}/p_i$. Note that $g = 1$ initially. After K such random walks, hence K values $g_1, \ldots, g_K$ of g, the estimate for the ith component of the solution vector is

$$\tilde{x}_i^{(K)} = x_i^{(0)} + \frac{(g_1 + g_2 + \cdots + g_K)}{K}.$$

The smallest K value required to achieve a given accuracy will be found in the next section and will be denoted K_0.

(b) Comparison of the effectiveness of a stationary linear iteration, the analogous Monte-Carlo process and Gaussian elimination. (*See* [55].) Given $\epsilon > 0$ and $x^{(0)}$, define the following quantities:

$$\delta = \frac{\epsilon}{\|r^{(0)}\|}$$

$$\alpha = \left[\frac{\log \delta + \log (1 - \|T\|)}{\log \|T\|}\right]$$

$$\beta = \left(\frac{1}{1 - \|T\|}\right)\left\{1 + \left[\frac{2}{\delta^2(1 - \|T\|)^2}\right]\right\},$$

where the brackets in α and β mean "largest integer in."

1. In the case of the stationary linear iteration we wish to find the smallest integer k such that

$$\|h - x^{(k)}\| < \epsilon,$$

or, the smallest k such that

$$\frac{\|T\|^k}{1 - \|T\|} < \delta.$$

This value of k is k_0, where $k_0 = 1 + \alpha$. The total number of multiplications required (a priori) to achieve $\|h - x^{(k)}\| < \epsilon$ is then

$$m = (k_0 - 1)n^2 + n + n^2 = n^2 + n + \alpha n^2,$$

where n is the order of the matrix T.

2. In the Monte-Carlo case we wish to find the smallest integer K such that

$$|h_i - \tilde{x}_i^{(K)}| < \epsilon.$$

Actually we find the smallest integer K such that the probability that $|h_i - \tilde{x}_i^{(K)}| < \epsilon$ is at least 95.45%. This value of K is

$$K_0 = \left[\frac{4\sigma^2(g:i_0 = i)}{\epsilon^2}\right] + 1,$$

where again the brackets mean "largest integer in." Hence the *average* number of multiplications required to assure (a priori) that $|h_i - \tilde{x}_i^{(K)}| < \epsilon$ with probability $> 95.45\%$ is

$$\bar{m} = n^2 + n + K_0 E(N:i_0 = i),$$

or, using the bounds for $\sigma^2(g:i_0 = i)$ and $E(N:i_0 = i)$,

$$\bar{m} = n^2 + n + \beta.$$

The value of K_0 may be estimated as

$$K_0 = (1 - \|T\|)\beta.$$

3. For Gaussian elimination, the number of multiplications needed to find the solution vector or any component of it is $m_g = \frac{1}{3}n^3 + n^2$. The inequalities that follow and Table 3.1 show the relations between m, $\bar{m}$, and m_g and the

TABLE 3.1

Favorable ranges of matrix order n for particular values of $\|T\|$ and δ

Method	$\|T\| = 0.5$, $\delta = 0.001$	$\|T\| = 0.9$, $\delta = 0.1$	$\|T\| = 0.9$, $\delta = 0.001$	$\|T\| = 0.9$, $\delta = 10^{-10}$
Gaussian elimination	$n \leq 33$	$n \leq 84$	$n \leq 261$	$n \leq 720$
Stationary linear	$34 \leq n \leq 1206$	For $n \leq 68$, $m < \bar{m}$ For $n \geq 130$, $m < m_g$	$262 \leq n \leq 4794$	$n \geq 721$
Monte-Carlo	$n \geq 1207$	$n \geq 85$	$n \geq 4795$	$n > 10^{10}$ approx

favorable ranges of n for each method:

$$m_g < m \quad \text{for} \quad 1 \leq n < \tfrac{1}{2}[3\alpha + (9\alpha^2 + 12)^{1/2}]$$

$$m < m_g \quad \text{for} \quad n > \tfrac{1}{2}[3\alpha + (9\alpha^2 + 12)^{1/2}]$$

$$\bar{m} < m_g \quad \text{for} \quad n > \tfrac{1}{2}[3\alpha + (9\alpha^2 + 12)^{1/2}]$$

$$m < \bar{m} \quad \text{for} \quad 1 \leq n < \left(\frac{\beta}{\alpha}\right)^{1/2}$$

$$m > \bar{m} \quad \text{for} \quad n > \left(\frac{\beta}{\alpha}\right)^{1/2}$$

(c) ***Monte-Carlo methods for the inverse matrix.*** Monte-Carlo procedures similar to the one described in (a) are available for computing an element of the inverse matrix, (see [55, 87]). The number of multiplications required is

$$m \sim n^3 \quad \text{for stationary linear iterations,}$$

$$\bar{m} \sim n^3 \quad \text{for Monte-Carlo, and}$$

$$m_g \sim \tfrac{1}{3}n^3 \quad \text{for Gaussian elimination.}$$

It should be noted that in any of these Monte-Carlo methods the work factor corresponding to the statistical portion of the method is independent of the matrix order, n.

3.3.2 Block-Iterative (Group, Implicit) Methods

3.3.2.1 General Discussion

In block iterative schemes several unknowns are connected together in the iteration formula in such a way that a linear subsystem must be solved before any one of them can be determined. The equations are divided into groups and the subsystem of equations belonging to a given group is solved for the corresponding unknowns using approximate values for the other unknowns.

For example, suppose the unknowns are divided into N groups so that $x_1, \ldots, x_{m_1}$ belong to group X_1; $x_{m_1+1}, \ldots, x_{m_2}$ belong to group X_2; etc. In general, $x_{m_{k-1}+1}, \ldots, x_{m_k}$ belong to group X_k. The matrix A is similarly divided into blocks A_{ij}, where the submatrix A_{ij} has m_i rows and $(m_j - m_{j-1})$ columns, and the b vector is divided into N groups $\beta_1, \ldots, \beta_N$. Then the system $Ax = b$ can be written

$$\begin{bmatrix} A_{11} & A_{12} & \cdots & A_{1N} \\ A_{21} & A_{22} & \cdots & A_{2N} \\ \cdot & \cdot & \cdots & \cdot \\ \cdot & \cdot & \cdots & \cdot \\ \cdot & \cdot & \cdots & \cdot \\ A_{N1} & A_{N2} & \cdots & A_{NN} \end{bmatrix} \begin{bmatrix} X_1 \\ X_2 \\ \cdot \\ \cdot \\ \cdot \\ X_N \end{bmatrix} = \begin{bmatrix} \beta_1 \\ \beta_2 \\ \cdot \\ \cdot \\ \cdot \\ \beta_N \end{bmatrix}.$$

Naturally the blocks are chosen so that solving each subsystem is as simple as possible.

Many of the point iterative methods in Section 3.3.1 can be generalized to block-iterative schemes. The Gauss–Seidel iteration [Section 3.3.1.1(b)] for example, when written as a block-iterative method becomes

$$X_i^{(k+1)} = A_{ii}^{-1}\left[\beta_i - \sum_{j=i+1}^{N} A_{ij}X_j^{(k)} - \sum_{j=1}^{i-1} A_{ij}X_j^{(k+1)}\right], \qquad i = 1, \ldots, N.$$

The only point-iterative method from Section 3.3.1 which will be expounded upon as a block iteration is the method of successive overrelaxation. The other block-iterative methods considered are the alternating-direction-implicit schemes, which are particularly effective when dealing with linear systems arising from elliptic partial differential equations.

Consider a second-order partial differential equation of elliptic type, as for example Laplace's equation, $\partial^2 x/\partial u^2 + \partial^2 x/\partial v^2 = 0$. It is defined over some region R completely enclosed by a boundary such that values of the function $x(u, v)$ are known on the boundary and are to be determined throughout R in such a way that the partial differential equation is satisfied. A rectangular mesh of width δ is placed over R, then the partial differential equation is approximated by some linear finite difference equation which must be satisfied for each point of the mesh. This difference equation usually involves $x(u, v)$ and a few of its neighboring points, $x(u + \delta, v + \delta)$, or $x(u, v + \delta)$, or $x(u + \delta, v)$. Assume some ordering of the points $x(u, v)$. Then each mesh point determines a linear equation and each equation contains only a few unknowns. Thus the resulting matrix is sparse. In addition this matrix often has other special properties that may be utilized in developing an iterative solution, for example, property A, p-cyclic, symmetric, positive definite. Since the matrix is also usually very large, the system is solved iteratively starting with initial guesses for $x(u, v)$ on the interior points of R. For a mesh of width $\delta < 1$ placed over the unit square, the number of interior points, hence the number of resulting linear equations is $n = (1/\delta - 1)^2$. The terms line, column, or row iteration are then special cases of block iteration which are meaningful in the context of the situation outlined above.

3.3.2.2 *Alternating-Direction Implicit (ADI) Methods* [12, 40, 50, 113]

Let $A = A_1 + A_2 + D$ where D is a nonnegative diagonal matrix, A_1 and A_2 are symmetric, positive definite, have positive diagonal elements and non-positive off-diagonal elements. Then the equation $Ax = b$ is equivalent to each of the equations

$$(A_1 + D + E_1)x = b - (A_2 - E_1)x,$$

$$(A_2 + D + E_2)x = b - (A_1 - E_2)x,$$

provided the matrices $(A_1 + D + E_1)$ and $(A_2 + D + E_2)$ are nonsingular. The two equations provide the basis for the alternating direction methods in which an intermediate vector $x^{(k+1/2)}$ is calculated from $x^{(k)}$ and in turn used to compute $x^{(k+1)}$. Two versions of alternating-direction will be considered, the Peaceman–Rachford method and the Douglas–Rachford method.

(a) Peaceman–Rachford. Letting $E_1^{(k)}$ be $\omega_k I$ and $E_2^{(k)}$ be $\tilde{\omega}_k I$, the two equations above yield the iteration formulas

$$(A_1 + D + \omega_k I)x^{(k+1/2)} = b - (A_2 - \omega_k I)x^{(k)},$$

and

$$(A_2 + D + \tilde{\omega}_k I)x^{(k+1)} = b - (A_1 - \tilde{\omega}_k I)x^{(k+1/2)},$$

or

$$x^{(k+1/2)} = [A_1 + D + \omega_k I]^{-1}[b - (A_2 - \omega_k I)x^{(k)}],$$

and

$$x^{(k+1)} = [A_2 + D + \tilde{\omega}_k I]^{-1}[b - (A_1 - \tilde{\omega}_k I)x^{(k+1/2)}],$$

which define the Peaceman–Rachford ADI method. Each equation can be rapidly solved, say by a Gaussian elimination, provided the matrices which have to be inverted are similar (by permutation matrices) to positive definite, well-conditioned tridiagonal matrices. The parameters ω_k, $\tilde{\omega}_k$ are chosen so as to accelerate convergence.

STATIONARY CASE Assume now that the process is stationary, that is, that $\omega_k = \omega$ and $\tilde{\omega}_k = \tilde{\omega}$. Letting $\omega = \tilde{\omega}$, the iteration matrix for the stationary Peaceman–Rachford method is

$$T_\omega = [A_2 + D + \omega I]^{-1}[A_1 - \omega I][A_1 + D + \omega I]^{-1}[A_2 - \omega I].$$

The rate of convergence is determined by the largest absolute eigenvalue of the error reduction matrix and must be less than 1 for convergence. Suggestions for choosing ω may be obtained by taking $k = 1$ in the nonstationary case. The following theorems are available for stationary ADI methods:

THEOREM 1 (WACHSPRESS AND HABETLER): The Peaceman–Rachford method is always convergent for $E_1 = E_2$, $(\omega = \tilde{\omega})$, when $(\omega I + \frac{1}{2}D)$ is positive definite and symmetric and $(A_1 + A_2 + D)$ is positive definite.

THEOREM 2 (WACHSPRESS, SHELDON, AND HABETLER): Any stationary *ADI* process with $E_1^{(k)} = E_1$, $E_2^{(k)} = E_2$ is convergent provided $(D + E_1 + E_2)$ is symmetric and positive definite, and $(2A_1 + D + E_1 - E_2)$ and $(2A_2 + D + E_2 - E_1)$ are positive definite. Let $A_3 = A_1 + \frac{1}{2}D$ and $A_4 = A_2 + \frac{1}{2}D$.

THEOREM 3: Let A_3 and A_4 be $n \times n$ Hermitian nonnegative definite matrices, where at least one of the matrices A_3 and A_4 is positive definite. Then for any $\omega > 0$, the Peaceman–Rachford method is convergent for any initial vector $x^{(0)}$.

NONSTATIONARY CASE Let $A = A_1 + A_2 + D$ and $Ax = b$, where D is a nonnegative diagonal matrix and A is nonsingular. Make the following assumptions:

$$A_1A_2 = A_2A_1,$$

$D = \sigma I$, σ a constant ≥ 0, A_1 and A_2 are similar to nonnegative diagonal matrices.

THEOREM 1: If A_1 and A_2 are positive definite symmetric and if $(A_1 + A_2)$ is irreducible, then the three assumptions above are equivalent to $A_1A_2 = A_2A_1$, $A_1D = DA_1$, $A_2D = DA_2$.

Now let $A_3 = A_1 + \frac{1}{2}D = A_1 + (\frac{1}{2}\sigma)I$, and $A_4 = A_2 + \frac{1}{2}D = A_2 + (\frac{1}{2}\sigma)I$. The iteration

$$(A_3 + \omega_k I)x^{(k+1/2)} = b - (A_4 - \omega_k I)x^{(k)}$$

$$(A_4 + \omega_k I)x^{(k+1)} = b - (A_3 - \omega_k I)x^{(k+1/2)}$$

defines a nonstationary Peaceman–Rachford method with iteration matrix

$$T^{(k)} = (A_4 + \omega_k I)^{-1}(A_3 - \omega_k I)(A_3 + \omega_k I)^{-1}(A_4 - \omega_k I).$$

Since

$$e^{(k+1)} = \prod_{i=0}^{k} T^{(i)} e^{(0)},$$

the average rate of convergence for fixed k is defined as

$$R_k = -\frac{1}{k} \log \left[\lambda \left(\prod_{i=0}^{k-1} T^{(i)} \right) \right]$$

where

$$\lambda \left(\prod_{i=0}^{k-1} T^{(i)} \right)$$

is the spectral radius or largest absolute eigenvalue of the matrix product,

$$M_k = \prod_{i=0}^{k-1} T^{(i)}.$$

Let μ_3 denote an eigenvalue of A_3, and μ_4 an eigenvalue of A_4. Choose α and β so that $\alpha \leq \mu_3, \mu_4 \leq \beta$ and let $c = \alpha/\beta$. Although optimal iteration parameters ω_k are not available, good ones that are simple to compute have been proposed. Two such choices for the ω_i are the following:

Peaceman–Rachford parameters. Let $\omega_{i-1} = \beta c^{(2i-1)/(2k)}$ for $i = 1, \ldots, k$. In the special case of a stationary iteration, $k = 1$ and $\omega = \sqrt{\beta\alpha}$. For fixed k and using the Peaceman–Rachford values for ω_i, it can be shown that

$$\lambda \left(\prod_{i=0}^{k-1} T^{(i)} \right) \leq \left[\frac{1 - c^{(1/2k)}}{1 + c^{(1/2k)}} \right]^2 = \delta^2$$

so that $R_k \geq -(2/k) \log \delta$. Writing k in terms of δ and equating the first derivative to zero gives the result that R_k is maximized when $\delta = \sqrt{2} - 1$. The procedure for computing parameters and convergence rate is

1. Estimate α and β, see Appendix B, Theorem 34.
2. Compute $c = \alpha/\beta$.
3. Find the smallest integer k such that $(\sqrt{2} - 1)^{2k} \leq c$.
4. Find ω_{i-1} from the formula above for $i = 1, \ldots, k$.
5. Find δ and $R_k \cong -(2/k) \log \delta$.

Wachspress parameters. Let $\omega_{i-1} = \beta c^{(i-1)/(k-1)}$ for $k \geq 2$, $i = 1, \ldots, k$. For fixed k and using the Wachspress parameters, it can be shown that

$$\lambda \left(\prod_{i=0}^{k-1} T^{(i)} \right) \leq \left[\frac{1 - c^{(1/2(k-1))}}{1 + c^{(1/2(k-1))}} \right]^2 = \epsilon^2,$$

so that $R_k \geq -(2/k) \log \epsilon$. The corresponding computation procedure is

1. Estimate α and β, see Appendix B, Theorem 34.
2. Compute $c = \alpha/\beta$.
3. Find the smallest integer k such that $(\sqrt{2} - 1)^{2(k-1)} \leq c$.
4. Find ω_{i-1} from the formula above for $k \geq 2$, $i = 1, \ldots, k$.
5. Find ϵ and $R_k \cong -(2/k) \log \epsilon$.

Other choices of iteration parameter are, of course, possible, for example those defined by de Boor and Rice (see [40]).

Numerical experiments [50] indicate that the Wachspress parameters are superior to the Peaceman–Rachford parameters by a factor of about 2, provided k is chosen as described. If μ_3, μ_4 are eigenvalues of A_3, A_4, respectively, corresponding to the same eigenvector, then

$$\frac{(\omega_k - \mu_3)(\omega_k - \mu_4)}{(\omega_k + \mu_3)(\omega_k + \mu_4)}$$

is the corresponding eigenvalue of the iteration matrix T_k. For any positive ω_k, the largest absolute eigenvalue of $T^{(k)}$ is less than 1. The sequence of parameters $\omega_0, \omega_1, \ldots, \omega_{k-1}$ gives eigenvalues

$$\lambda = \prod_{i=0}^{k-1} \frac{(\omega_i - \mu_3)(\omega_i - \mu_4)}{(\omega_i + \mu_3)(\omega_i + \mu_4)}$$

of the product matrix

$$\prod_{i=0}^{k-1} T^{(i)}.$$

For a given k, optimal ω_i's would be those which minimize the maximum absolute value of the product λ over all μ_3, μ_4 of A_3 and A_4, respectively. No simple formulas for optimal parameters can be given.

Using either set of parameters, an estimate for the factor of reduction of the error after k double sweeps is

$$P_k = \left[\left(\frac{1 - \sqrt{x}}{1 + \sqrt{x}}\right) e^{-x^{3/2}/(1-x)}\right]^4$$

where $x = c^{1/(k-1)}$ for small c. The integer k is chosen so that $S_k = (P_k)^{1/k}$, the average factor of reduction for the double sweep, is minimized.

(b) Douglas–Rachford. The first version of the Douglas–Rachford method is given by letting $E_1^{(k)} = E_2^{(k)} = \omega_k I - \frac{1}{2}D$. Thus

$$x^{(k+1/2)} = (A_3 + \omega_k I)^{-1}(b - (A_4 - \omega_k I)x^{(k)})$$

$$x^{(k+1)} = (A_4 + \omega_k I)^{-1}(A_4 x^{(k)} + \omega_k x^{(k+1/2)}),$$

where $A_3 = A_1 + \frac{1}{2}D$ and $A_4 = A_2 + \frac{1}{2}D$.

An alternate version takes $E_1^{(k)} = E_2^{(k)} = \omega_k I$, to give

$$x^{(k+\frac{1}{2})} = (A_1 + D + \omega_k I)^{-1}(b - (A_2 - \omega_k I)x^{(k)})$$
$$x^{(k+1)} = (A_2 + D + \omega_k I)^{-1}[(A_2 + \tfrac{1}{2}D)x^{(k)} + (\tfrac{1}{2}D + \omega_k I)x^{(k+\frac{1}{2})}].$$

The iteration or error reduction matrix for the first version is

$$T_k = (A_4 + \omega_k I)^{-1}(A_3 + \omega_k I)^{-1}(A_3 A_4 + \omega_k{}^2 I)$$
$$= [A_3 A_4 + \omega_k(A_4 + A_3) + \omega_k{}^2 I]^{-1}(A_3 A_4 + \omega_k{}^2 I)$$

and for the second version is

$$T_k = (A_2 + D + \omega_k I)^{-1}[(\tfrac{1}{2}D + \omega_k I)(A_1 + D + \omega_k I)^{-1}(\omega_k I - A_2) + (A_2 + \tfrac{1}{2}D)].$$

For the stationary case the Theorem 2 of section (a) (the Wachspress, Sheldon, and Habetler theorem) and the last theorem of that section hold for the Douglas–Rachford method.

The assumptions for the nonstationary case are made in the preceding section. Of the two Douglas–Rachford methods it can be shown that the first version is superior; hence it will be the only one considered in the remarks that follow.

For fixed k and assuming that the Peaceman–Rachford parameters [see Section (a)] are used in the first Douglas–Rachford method, the spectral radius of the product of the first k iteration matrices satisfies

$$\lambda\left(\prod_{i=0}^{k-1} T^{(i)}\right) \leq \tfrac{1}{2}(1 + \delta^2),$$

where

$$\delta = \left(\frac{1 - c^{1/2k}}{1 + c^{1/2k}}\right)$$

as defined for the Peaceman–Rachford parameters, and $T^{(i)}$ is the ith iteration matrix for the first version of the Douglas–Rachford method.

An estimate of c, an optimal integer k, parameters ω_i, and the average convergence rate for this k value are found as follows:

1. Estimate α and β.
2. Compute $c = \alpha/\beta$.
3. Find the smallest integer k so that $(0.25)^{2k} \leq c$.
4. Find ω_{i-1} from the formula for the Peaceman–Rachford parameters:

$$\omega_{i-1} = \beta c^{(2i-1)/(2k)}, \qquad i = 1, \ldots, k.$$

5. Find δ and

$$R_k \cong -\frac{1}{k}\log\left[\tfrac{1}{2}(1+\delta^2)\right]$$

Note. This Douglas–Rachford method with Peaceman–Rachford parameters is much less effective than the Peaceman–Rachford method with these parameters for fixed k, or for k chosen as a function of c. The Douglas–Rachford method is even more inferior to Peaceman–Rachford when Wachspress parameters are used for both, and k depends upon c as previously defined. This does *not necessarily* imply that the Douglas–Rachford method would be inferior to the Peaceman–Rachford one if *optimal* parameters were used.

3.3.2.3 *Block Successive Overrelaxation* [40, 44]

Assume that A has been subdivided into blocks A_{ij} as described in Section 3.3.2.1. Let $D = (A_{ii})$, $i = 1, \ldots, N$; let $U = (A_{ij})$, $j > i$; and let $L = (A_{ij})$, $j < i$. Then block successive overrelaxation is defined by

$$X_i^{(k+1)} = -\omega\left[\sum_{j=1}^{i-1} A_{ii}^{-1}A_{ij}X_j^{(k+1)} + \sum_{j=i+1}^{N} A_{ii}^{-1}A_{ij}X_j^{(k)} - A_{ii}^{-1}\beta_i\right] - (\omega - 1)X_i^{(k)}$$

for $i = 1, \ldots, N$. In matrix notation this is

$$X^{(k+1)} = -\omega D^{-1}[LX^{(k+1)} + UX^{(k)} - \beta] - (\omega - 1)IX^{(k)},$$

where $A = D + L + U$.

THEOREM 1: If A is symmetric and A_{ii} is positive definite for each $i = 1, \ldots, N$, then the method of block successive overrelaxation converges for $\omega = 1$ (i.e., the block Seidel method) if and only if A is positive definite.

THEOREM 2: If A is Hermitian and D is Hermitian positive definite, then the method of block successive overrelaxation converges for all $X^{(0)}$ if and only if $0 < \omega < 2$ and A is positive definite.

The optimal relaxation factor is $\omega_{\text{opt}} = 2/[1 + (1 - \lambda^2)^{1/2}]$, where λ is the largest absolute eigenvalue of the matrix $(-D^{-1}A + I)$, the block Jacobi iteration matrix. Also, the rate of convergence of the block SOR method with optimal ω is about twice the square root of the rate of convergence of the block Seidel method.

An advantage is always obtained by using *block* SOR rather than *point* SOR whenever the sufficient conditions

1. $|A| \neq 0$
2. $\sum_{\substack{j=1 \\ j\neq i}}^{n} |a_{ij}| \leq a_{ii}$
3. $a_{ij} \leq 0$ for $i \neq j$

are satisfied. This is not true in general, even for positive-definite matrices. Of particular interest are the "two-line" iterations, $N = 2$, suggested by the situation described earlier with respect to partial differential equations. The Chebyshev accelerations of Section 3.4.4 have also been applied to obtain varying ω's, hence nonstationary, though cyclic, Chebyshev variants.

3.4 ACCELERATION PROCEDURES AND IMPROVED SOLUTIONS

3.4.1 Aitken's Δ^2 Method [17, 22, 38]

Given a sequence of real or complex numbers $\{x_n\}$ that converges linearly to a limit α, Aitken's method produces a new sequence $\{y_n\}$ that converges more rapidly to α.

Let $e_i = \alpha - x_i$. The sequence $\{x_n\}$ is said to have *geometric convergence* if $e_{i+1} = Ke_i$, where K is a constant, $|K| < 1$. The sequence $\{x_n\}$ is said to have *linear convergence* if $e_{i+1} = (K + \sigma_i)e_i$, where K is a constant, $|K| < 1$, and $\sigma_i \to 0$ as $i \to \infty$. Geometric convergence is then a special case of linear convergence in which all $\sigma_i \equiv 0$.

For the case of geometric convergence, K can be eliminated between successive relations to obtain

$$\alpha = \frac{x_{i+2}x_i - x_{i+1}^2}{x_{i+2} - 2x_{i+1} + x_i},$$

so that given any three successive terms of the sequence, the limit α can be calculated. This suggests that in the case of linear convergence an improvement in convergence might be obtained by defining an auxiliary sequence as follows. For a linearly converging sequence $\{x_i\}$ with limit α, define

$$y_i = \frac{x_{i+2}x_i - x_{i+1}^2}{x_{i+2} - 2x_{i+1} + x_i},$$

which may be written in any of the following three forms:

$$y_i = x_i - \frac{(x_{i+1} - x_i)^2}{x_{i+2} - 2x_{i+1} + x_i}$$

$$y_i = x_{i+1} - \frac{(x_{i+1} - x_i)(x_{i+2} - x_{i+1})}{x_{i+2} - 2x_{i+1} + x_i}$$

$$y_i = x_{i+2} - \frac{(x_{i+2} - x_{i+1})^2}{x_{i+2} - 2x_{i+1} + x_i}.$$

This transformation is called the Aitken Δ^2 method. The last expression for y_i is probably best for computational purposes. If $t_i = \alpha - y_i$ and $e_i = \alpha - x_i$, then it can be shown that for i sufficiently large $t_i/e_i \to 0$; hence that the y sequence converges faster than the x sequence.

Other acceleration procedures include the Shank's transform, Euler summation process, Richardson's h^2 extrapolation, but these are not discussed here.

3.4.2 Method of Hotelling and Bodewig [19]

The method of Hotelling and Bodewig is useful for improving an estimated solution obtained by some other means, for example, by the Seidel iteration or by a Monte-Carlo process.

Assume $Ax = b$ has an approximate solution x_0 to its true solution $A^{-1}b$, or an approximate inverse C_0 to A^{-1}.

Then $x_1 = x_0 + C_0 r = C_0(2I - AC_0)b$ is a better approximation to $x = A^{-1}b$ than x_0, and $C_1 = C_0(2I - AC_0)$ is a better approximation to A^{-1} than C_0. Let $B_i = I - AC_i$, where $C_{i+1} = C_i(I + B_i)$. Then $B_{i+1} = B_i^2$, so that $B_k = B_0^{2k}$. Since $B_0 = I - AC_0$,

$$AC_0 = I - B_0$$

$$A^{-1} = C_0(I - B_0)^{-1} = C_0(I + B_0 + B_0^2 + \cdots),$$

which converges if the eigenvalues of B_0 are less than 1 in absolute value. It may be shown that the foregoing expression may be written

$$A^{-1} = C_0(I + B_0)(I + B_0^2)(I + B_0^4) \cdots .$$

This last expression describes the procedure of Hotelling and Bodewig and employs even powers of the matrix B_0. Unfortunately, the method requires matrix multiplications, which are time consuming if very many are needed for the desired accuracy.

3.4.3 Iterated Direct Methods [7, 60]

Suppose that the system $Ax = b$ has been solved by a direct method and the solution $x^{(1)}$ has been obtained. Can this solution now be improved?

Let $r^{(1)} = b - Ax^{(1)}$ be the residual vector (or matrix). Solve $A(\Delta x^{(1)}) = r^{(1)}$ for $\Delta x^{(1)}$. (If an approximate inverse is available, this solution requires little extra work.) Let $x^{(2)} = x^{(1)} + \Delta x^{(1)}$, $r^{(2)} = b - Ax^{(2)}$ and solve $A(\Delta x^{(2)}) = r^{(2)}$ for $\Delta x^{(2)}$, etc. Denote by $\tilde{A}^{-1}$ the approximate value of A^{-1}, so that

$$x^{(1)} = \tilde{A}^{-1}b,$$

$$\Delta x^{(1)} = \tilde{A}^{-1}r^{(1)}, \text{ etc.}$$

Then

$$\begin{aligned} r^{(1)} &= b - A(\tilde{A}^{-1}b) = (I - A\tilde{A}^{-1})b \\ r^{(2)} &= b - A(x^{(1)} + \Delta x^{(1)}) = (I - A\tilde{A}^{-1})b - A\,\Delta x^{(1)} \\ &= (I - A\tilde{A}^{-1})b - A\tilde{A}^{-1}r^{(1)} = (I - A\tilde{A}^{-1})b - A\tilde{A}^{-1}(I - A\tilde{A}^{-1})b \\ &= (I - A\tilde{A}^{-1})^2 b \end{aligned}$$

and, in general

$$r^{(k)} = (I - A\tilde{A}^{-1})^k b.$$

Now $r^{(k)} \to 0$ as $k \to \infty$ if and only if all of the eigenvalues of $(I - A\tilde{A}^{-1})$ are less than 1 in absolute value. The convergence is linear and

$$x^{(k)} = x^{(1)} + \Delta x^{(1)} + \Delta x^{(2)} + \cdots + \Delta x^{(k)} \to x.$$

3.4.4 Linear Chebyshev Accelerations for Linear Stationary Processes [93]

3.4.4.1 First-Degree Chebyshev Accelerations

Let $x^{(k+1)} = (I - CA)x^{(k)} + Cb$ define a linear stationary iteration for solving $Ax = b$. Consider the process

$$\begin{aligned} p^{(k+1)} &= (I - CA)x^{(k)} + Cb \\ x^{(k+1)} &= p^{(k+1)} + \omega_k(p^{(k+1)} - x^{(k)}) \\ &= (1 + \omega_k)p^{(k+1)} - \omega_k x^{(k)}, \qquad k = 1, 2, \ldots, \end{aligned}$$

which defines a linear acceleration with acceleration factors ω_k. If $p^{(k+1)}$ is eliminated between the two equations, the resulting iteration

$$x^{(k+1)} = [I - (1 + \omega_k)CA]x^{(k)} + (1 + \omega_k)Cb$$

is again linear, though no longer stationary unless all $\omega_k = \omega$. [Note that the extrapolated Jacobi method of Section 3.3.1.2 may be obtained from the Jacobi method by applying the above linear acceleration with fixed ω, and that Richardson's method, Section 3.3.1.3, is a linear Chebyshev acceleration of the Jacobi method with varying ω_k.] The ith iteration matrix is

$$T^{(i)} = [I - (1 + \omega_i)CA],$$

and the product of the first k iteration matrices is

$$M_k = \prod_{i=0}^{k-1} T^{(i)}.$$

Now $M_k = M_k(CA)$ is a polynomial of degree k in CA such that $M_k(0) = 1$ and $M_k(1/[1 + \omega_i]) = 0$. We wish to choose ω_i so as to minimize the largest absolute eigenvalue of M_k. By Theorem 43 of Appendix B, the eigenvalues of $M_k(CA)$ are $M_k(\lambda_j)$, where λ_j is an eigenvalue of CA. Assuming CA is symmetric, we wish to minimize the maximum magnitude of the polynomial $M_k(x)$, $a' \le x \le b'$, where a' and b' are the minimum and maximum eigenvalues respectively of CA. Letting $y = (b' + a' - 2x)/(b' - a')$, we have $y_0 = y(0) = (b' + a')/(b' - a')$, $y(a') = 1$, $y(b') = -1$, and the desired polynomial is $M_k(x) = T_k(y)/T_k(y_0)$, where $T_k(y)$ is the Chebyshev polynomial of degree k over the interval $-1 \le y \le 1$. $M_k(x)$ is a Chebyshev polynomial of degree k over the interval $a' \le x \le b'$. It is known that $M_k(x)$ has maximum value $1/T_k(y_0)$ and has zeros $x_j^{(k)}$ corresponding to the roots $y_j^{(k)}$ of $T_k(y)$. The zeros of $T_k(y)$ are $y_j^{(k)} = \cos\,[(2j + 1)\pi/(2k)], j = 0, \ldots, k - 1$, so those of $M_k(x)$ are

$$x_j^{(k)} = \tfrac{1}{2}[(b' + a') - y_j^{(k)}(b' - a')], \qquad j = 0, \ldots, k - 1.$$

Hence $1/(1 + \omega_j) = x_j^{(k)}$ or $\omega_j = (1 - x_j^{(k)})/x_j^{(k)}, j = 0, \ldots, k - 1$, and the spectral radius of M_k is $\lambda(M_k) = 1/T_k(y_0)$, so that the average rate of convergence is

$$R(M_k) = -\frac{1}{k}\log\left(\frac{1}{T_k(y_0)}\right) = \frac{1}{k}\log T_k(y_0).$$

To perform this Chebyshev acceleration, one must choose k, estimate a' and b', then compute the ω_i factors. When $[2 - (b' + a')]/(b' - a')$ is close to 1 and k is large, some of the ω_i become large, thus possibly making the iteration unstable due to rounding error. In such cases a second-degree acceleration could be used.

3.4.4.2 Second-Degree Chebyshev Accelerations

Again let

$$y(x) = \frac{b' + a' - 2x}{b' - a'},$$

$$y_0 = y(0) = \frac{b' + a'}{b' - a'},$$

$$y_1 = y(1) = \frac{b' + a' - 2}{b' - a'}.$$

Define

$$\alpha_k = \frac{T_{k-1}(-y_1)}{T_k(-y_1)}$$

$$\beta_k = \frac{T_{k-2}(-y_1)}{T_k(-y_1)}$$

and the iteration

$$p^{(1)} = (I - CA)x^{(0)} + Cb$$

$$x^{(1)} = \frac{2\alpha_1 p^{(1)}}{b' - a'} - y_0\alpha_1 x^{(0)}$$

$$\vdots \qquad \vdots$$

$$p^{(k)} = (I - CA)x^{(k-1)} + Cb$$

$$x^{(k)} = \frac{4\alpha_k p^{(k)}}{b' - a'} - 2y_0\alpha_k x^{(k-1)} - \beta_k x^{(k-2)}, \qquad k > 1.$$

This iteration defines a second-degree acceleration. Eliminating $p^{(k)}$ between the last two equations gives

$$x^{(k)} = 2\alpha_k\left[\left(\frac{2}{b' - a'}\right)(I - CA) - y_0\right]x^{(k-1)} - \beta_k x^{(k-2)} + \left(\frac{4}{b' - a'}\right)\alpha_k Cb,$$

which is now a second-degree linear iteration and non-stationary. The iteration matrix is $M_k(x) = T_k(-y)/T_k(-y_1)$ for every k, not just for preassigned k as in the previous first-degree case. To perform this acceleration, a' and b' must be estimated and then

$$-y_1 = \frac{2 - (b' + a')}{b' - a'}$$

computed. Then the Chebyshev polynomials must be evaluated at $y = -y_1$ for each k. This may be done recursively as follows:

$$T_0(y) = 1$$

$$T_1(y) = y$$

$$\vdots \qquad \vdots$$

$$T_k(y) = 2yT_{k-1}(y) - T_{k-2}(y), \qquad k > 1.$$

The α_k and β_k may then be computed since they are ratios of the Chebyshev polynomials. The α_k and β_k remain bounded and approach constant values as $k \to \infty$. A disadvantage of this second-degree acceleration is that storage for the iterate $x^{(k-2)}$ is required in addition to that for $x^{(k-1)}$.

4

ILL-CONDITIONING— MEASURES OF CONDITION*

When small errors in the coefficients or right-hand sides or in the process of solution of a system of linear equations have little effect on the solution, the system is termed "well conditioned"; if the effect is large, that is, the solution is very sensitive to the values of the coefficients, the problem is "ill conditioned." To say that a matrix is ill conditioned, we must specify with respect to what operation. A matrix may be ill conditioned with respect to solving a system of equations but not with respect to finding eigenvalues. All methods of solving an ill-conditioned system of linear equations are generally bad; a direct method produces significant roundoff error and loses significant figures in the pivots; an iterative method converges very slowly.

Ill conditioning with respect to linear systems occurs when the matrix is nearly singular, that is, when some of the rows are almost linearly dependent or when some of the hyperplanes defined by the equations are approximately parallel. However, the use of $|A|$, the determinant of A, as a measure of condition is not quite valid since the equations can be multiplied by a constant to obtain any value of $|A|$. When the equations are normalized, however, by dividing the ith row by

$$\left(\sum_{j=1}^{n} a_{ij}^{2}\right)^{1/2},$$

then the determinant of A may be used as a test for ill condition. If the determinant of the normalized A is small compared with ± 1, the system is ill conditioned.

* See references [8, 13, 27, 41, 42, 85, 98–101].

As an example, take the system

$$x_1 + 10x_2 = 11$$
$$10x_1 + 101x_2 = 111,$$

whose correct solution is $x_1 = x_2 = 1$. The determinant of the coefficient matrix A is $|A| = 1$. The eigenvalues of A are $\lambda_1 = 101.9902$ and $\lambda_2 = 0.0098$. To normalize the system, divide the first equation by $\sqrt{101}$ and the second by $\sqrt{10301}$. Then $|A| \sim 0.001$ for the normalized system.

For an ill-conditioned system, a set of values for the unknowns which differs considerably from the solution may give small residuals. Since the error vector is $e = A^{-1}r$, a small residual r does not necessarily imply small e. For the system above, the set of values

$$x_1 = 1.001, \quad x_2 = 1.01 \quad \text{gives} \quad r_1 = 0.01, \quad r_2 = 0.01, \quad \sum_{i=1}^{2} r_i^2 = 0.0002,$$

whereas the set of values

$$x_1 = 11.1, \qquad x_2 = 0 \quad \text{gives} \quad r_1 = 0.01, \qquad r_2 = 0, \qquad \sum_{i=1}^{2} r_i^2 = 0.0001.$$

The second set has smaller residuals, but the first set is closer to the true solution, $x_1 = x_2 = 1$.

Hence a residual of zero implies the solution is exact, but a residual vector that is small does not necessarily imply the solution is close. A small residual means there are no substantial errors in the direction of the large eigenvectors, but there may be very substantial errors in the direction of the small eigenvectors. When a matrix has all its eigenvalues close together so that $|\lambda_{\max}/\lambda_{\min}| \sim 1$, the matrix is well conditioned for solving linear equations (but very ill conditioned for determining eigenvalues). For the preceding example

$$\left|\frac{\lambda_{\max}}{\lambda_{\min}}\right| = \frac{101.9902}{0.0098} \sim 10{,}407.$$

The *condition number* is a measure of the effect on an approximate inverse of a matrix A (or on an approximate solution of $Ax = b$), when the calculations are carried to a fixed number of places and the elements of A are changed slightly.

The following measures of condition have been suggested for the $n \times n$ matrix, A:

1. Turing's M-condition number, $M(A)$:

$$M(A) = n \max_{i,j} |a_{ij}| \max_{i,j} |\alpha_{ij}|,$$

where $A = (a_{ij})$ is an $n \times n$ matrix and $A^{-1} = (\alpha_{ij})$ is its inverse. An average value for $M(A)$ is $\sqrt{n} \log n$.

2. Turing's N-condition number, $N(A)$:

$$N(A) = n^{-1} \|A\| \, \|A^{-1}\|, \quad \text{where} \quad \|A\| = \left(\sum_{i,j} a_{ij}^2\right)^{1/2}.$$

An average value for $N(A)$ is $\sqrt{n}$.

3. Von Neumann and Goldstine's P-condition number, $P(A)$:

$$P(A) = \left|\frac{\lambda_{\max}}{\lambda_{\min}}\right|,$$

where $\lambda_{\max}$ and $\lambda_{\min}$ are the largest and the smallest respectively (in absolute value) of the eigenvalues of A. An average value for $P(A)$ is n.

4. Determinant of the normalized matrix. Divide the ith row of

$$A = (a_{ij}) \quad \text{by} \quad \left(\sum_{j=1}^{n} a_{ij}^2\right)^{1/2}$$

to obtain a normalized matrix. If the determinant of the normalized A is small compared with ± 1, the system is ill conditioned.

General relations which hold between the first three measures are

$$\frac{M(A)}{n^2} \leq N(A) \leq M(A)$$

$$\frac{M(A)}{n} \leq P(A) \leq nM(A).$$

The best-conditioned matrices are the orthogonal matrices, which have N-condition numbers of 1 and M-condition numbers about $\log n$. If the coefficients of a matrix were chosen at random from a normal population, an N-condition number of order $\sqrt{n}$ and an M-condition number of order $\sqrt{n} \log n$ would be obtained.

Note from the theorem which follows that symmetrization of an ill-conditioned matrix cannot improve its condition. Apart from the extra work involved, this presents an objection to using $A'Ax = A'b$ to solve $Ax = b$.

THEOREM 1 (TAUSSKY): Let A be a real $n \times n$ nonsingular matrix and A' its transpose. Then AA' is more "ill conditioned" than A in the sense that $P(AA') \geq P(A)$ and $N(AA') \geq N(A)$.

The condition numbers given above are not of great practical use since they require knowledge of the inverse, or eigenvalues, or determinant of the matrix and usually these are just the quantities we wish to determine. However, some insight into the origin of the condition measures 1 and 2 may be gained from the following considerations.

Suppose that a change Δb in the right-hand sides of the system $Ax = b$ causes a change Δx in the solution. Then it can be shown that

$$\frac{\|\Delta x\|}{\|x\|} \leq \frac{\|A\| \, \|A^{-1}\| \, \|\Delta b\|}{\|b\|}.$$

Hence the quantity $\|A\| \, \|A^{-1}\|$ is a reasonable measure of condition.

Suppose that a change ΔA in the matrix A of the system $Ax = b$ causes a change Δx in the solution.

Then it may be shown that

$$\frac{\|\Delta x\|}{\|x\|} \leq \frac{\|A\| \, \|A^{-1}\| \dfrac{\|\Delta A\|}{\|A\|}}{1 - \|A\| \, \|A^{-1}\| \dfrac{\|\Delta A\|}{\|A\|}} \quad \text{provided} \quad \|A^{-1}\| \, \|\Delta A\| < 1.$$

5

ERROR MEASURES*

5.1 DEFINITIONS OF ERROR MEASURES

Given the system $AX = B$ whose solution is $X = A^{-1}B$, define an error matrix E and a residual matrix R from any appropriate solution $\tilde{X}$:

$$E = (e_{ij}) = (A^{-1}B - \tilde{X}),$$
$$R = (r_{ij}) = (B - A\tilde{X}).$$

When solving for A^{-1}, let $B = I$ so that

$$E = (e_{ij}) = (A^{-1} - \tilde{X}),$$
$$R = (r_{ij}) = (I - A\tilde{X}).$$

Now define the following three measures of error as applied to both the error and residual matrices:

1. $a_e = \dfrac{1}{n^2}\sum_{i,j} |e_{ij}|\ ; \qquad a_r = \dfrac{1}{n^2}\sum_{i,j} |r_{ij}|$

2. $f_e = \dfrac{1}{n}\left[\sum_{i,j} e_{ij}^2\right]^{1/2}; \qquad f_r = \dfrac{1}{n}\left[\sum_{i,j} r_{ij}^2\right]^{1/2}$

3. $m_e = \max\limits_{i,j} |e_{ij}|\ ; \qquad m_r = \max\limits_{i,j} |r_{ij}|\ .$

Then $a_e \leq f_e \leq m_e$ and $a_r \leq f_r \leq m_r$.

* See references [41, 42, 45, 68, 76, 85, 101, 108].

5.2 EXPRESSIONS FOR THE ERROR IN SPECIAL CASES

5.2.1 Von Neumann–Goldstine

Von Neumann and Goldstine have given expressions for the error involved in the Gaussian elimination process:

$$\|A\tilde{A}^{-1} - I\| \leq 14.24P(A)n^2 2^{-t}$$

when A is symmetric and positive definite,

$$\|A\tilde{A}^{-1} - I\| \leq 36.58[P(A)]^2 n^2 2^{-t}$$

when A is nonsingular, where

$$\|A\| = \max_{\|\xi\|=1} \|A\xi\|,$$

$\|\xi\|$ is the Euclidean vector norm, $\tilde{A}^{-1}$ is an approximation to A^{-1} obtained by elimination,

$$P(A) = \left|\frac{\lambda_{\max}}{\lambda_{\min}}\right|$$

as defined in Chapter 4, and t is the number of binary places carried in the computation.

5.2.2 Wilkinson

Wilkinson has given error expressions in the following cases where floating-point arithmetic is used:

Let t be the number of binary places carried in the mantissa for floating-point arithmetic. Let

$$\|A\|_S = \max\,[\lambda(AA')]^{1/2}, \qquad \|A\|_\infty = \max_i \sum_{j=1}^{n} |a_{ij}|\,.$$

1. Then for the set of upper triangular equations $UX = B$, where

$$\tfrac{1}{2} \leq \max_{i,j} |u_{ij}| \leq 1,$$

strict error bounds are given by

$$\|UX - B\|_S \leq 2^{-t}[\sqrt{n}\|B\|_S + 0.4n^{5/2}\,\|X\|_S],$$

$$\|UX - I\|_S \leq 0.4(2^{-t})n^{5/2}\,\|X\|_S,$$

and

$$\frac{\|X - U^{-1}\|_S \leq 0.4(2^{-t})n^{5/2}\ \|U^{-1}\|_S^{\,2}}{(1 - 0.4(2^{-t})n^{5/2}\ \|U^{-1}\|_S)}.$$

2. For the lower triangular system $Lx = b$, where $|1_{ij}| \leq g$,

$$\|b - Lx\|_\infty \leq 0.53(2^{-t})g(n^2 + n + 2)\ \|x\|_\infty.$$

3. For Gaussian elimination with complete positioning for size performed for general matrix inversion, let $AX = I$ and

$$R = \max \left[\frac{\max |a_{ij}^{(k)}|}{\max |a_{ij}^{(1)}|}\right] = \max \left(\frac{P_k}{p_1}\right)$$

for $k = 1, \ldots, n$, where p_k is the kth pivot chosen. If $|a_{ij}| \leq 1$, then

$$\|AX - I\|_S \leq 0.78(2^{-t})n^{7/2}R\ \|A^{-1}\|_S.$$

4. For solution of a general system of linear equations $Ax = b$ using Gaussian elimination with some form of pivoting, let

$$|a_{ij}^{(k)}| \leq g \quad \text{for} \quad \text{all} \quad k = 1, \ldots, n.$$

Then

$$\|b - Ax\|_\infty \leq 1.06g(2^{-t})(2.005n^2 + n^3 + 0.265n^4 2^{-t})\|x\|_\infty.$$

(Wilkinson has found that in practice

$$\|b - Ax\|_\infty \leq 1.06g(2^{-t})n\|x\|_\infty$$

is usually true.) Similarly, for matrix inversion,

$$\|AX - I\|_\infty \leq 1.06g(2^{-t})(2.005n^2 + n^3 + 0.265n^4 2^{-t})\ \|X\|_\infty.$$

5. Finally, for the Cholesky method for solving $AX = I$, A symmetric and positive definite, $\frac{1}{2} \leq |a_{ij}| \leq 1$, we have

$$\frac{\|X - A^{-1}\|_S}{\|A^{-1}\|_S} < n(2^{-t})[1.12\ \|A^{-1}\|_S + n(1.66\ \|A^{-1}\|_S^{1/2} + 1.3)].$$

5.3 INTERVAL ARITHMETIC

Another approach to error analysis is provided by the techniques of interval arithmetic. Let A^I be a matrix whose elements are all intervals. Let A_c be the matrix whose elements are the centers of the intervals in A^I. The object is to compute an interval matrix C^I containing A^{-1} for every A in A^I. Eldon Hansen [71] has observed that $(A^I)^{-1}$ can be written as $B(A^I B)^{-1}$ where

$A^I B = I - E^I$ and the elements of E^I are small. Then errors involved in inverting A^I by interval arithmetic are replaced by errors caused by inverting $A^I B$, which are considerably smaller.

First, a single-precision approximation B is computed for $(A_c)^{-1}$. Next a double-precision approximation for A^I and double-precision arithmetic are used to compute the interval matrix

$$E^I = I - A^I B,$$

which is then rounded appropriately to single-precision accuracy. An element of E^I is denoted by $[x_{ij}, y_{ij}]$. A single-precision computation of r is then performed, where r is defined as

$$r = \max_{1 \le i \le n} \sum_{j=1}^{n} [\max(|x_{ij}|, |y_{ij}|)].$$

We assume that r is less than unity. An integer k is chosen so that $b'(k) = r^{k+1}(1 - r)^{-1}$ is less than the minimum interval length in E^I, or, in the case where this minimum length is zero, choose k as the low-order bit of a representative element of E^I. Compute

$$G^I = E^I(I + E^I(I + E^I(I + \cdots + E^I)\cdots))$$

for k factors, in single precision and let $S^I = I + G^I$. Also let P^I be the interval matrix with any element equal to $[-b', b']$. Find $(G^I + P^I)$ and $B(G^I + P^I)$, then use double precision to add B. The result is finally $C^I = B + B(G^I + P^I)$.

5.4 REMARKS

The error bounds given in Section 5.2 are all rather severe upper bounds and are hardly ever realized in practice. The statistical distribution of rounding errors will reduce the function of n occurring in the error expressions, possibly to its square root, so that computed inverses are much more accurate than expected from the given bounds.

According to Wilkinson [41], for a very large class of triangular matrices the relative error involved in the solution of triangular sets of equations is not, in practice, at all dependent upon the condition number $\|L\| \cdot \|L^{-1}\|$. Even when L is ill conditioned, the solution of $Lx = b$, or of $LX = I$ is far more accurate than expected, and the relative error is commonly unaffected by the condition of the matrix. This is always true for triangular matrices having positive diagonal and negative off-diagonal elements.

For compact methods of triangular decomposition, especially when A is ill conditioned, the error in the triangular decomposition is more important for accuracy than the error in the solution of the triangular system. Partial pivoting should be used during the decomposition.

For a positive-definite symmetric matrix, A, which is decomposed as $A = LL' = U'U$, L and U real, if $|a_{ij}| \leq 1$, then $|l_{ij}| \leq 1$, so that the Cholesky method has a guaranteed stability and requires no pivoting.

If A is symmetric, but not positive definite and is decomposed triangularly so that $l_{rr} = 1$, then U differs from L' only in the signs of some of its rows. Stability is no longer guaranteed as in the positive-definite case and poor inverses may result even for well-conditioned matrices if pivoting is not used.

A small residual matrix $(A\tilde{X} - I)$ necessarily implies that $\tilde{X}$ is a good inverse, whereas a small residual vector $(A\tilde{x} - b)$ does not necessarily imply that $\tilde{x}$ is a good solution vector.

6

SCALING*

In general it is undesirable that in the system $Ax = b$ the elements of A (or b) be of different orders of magnitude. Uneven scaling can usually be eliminated by one of the following methods:

1. Divide each element of a column by the length of that column, that is, divide the jth column of A by

$$\left(\sum_{i=1}^{n} a_{ij}^2\right)^{1/2} = \frac{1}{t_j}$$

and the b column by

$$\left(\sum_{i=1}^{n} b_i^2\right)^{1/2} = \frac{1}{t_{n+1}}.$$

The solutions y_j of the transformed system are written in terms of the solutions x_j of the original system as follows:

Let $D = \text{diag}\,(t_1, \ldots, t_n)$. Then we can write $(AD)y = t_{n+1}b$,

$$y = (AD)^{-1}t_{n+1}b = (t_{n+1}D^{-1})A^{-1}b,$$

or, since $Ax = b$,

$$y = (t_{n+1}D^{-1})x, \quad \text{that is,} \quad y_j = \left(\frac{t_{n+1}}{t_j}\right)x_j,$$

and

$$x = \left(\frac{D}{t_{n+1}}\right)y, \quad \text{that is,} \quad x_j = \left(\frac{t_j}{t_{n+1}}\right)y_j.$$

Similarly, to find A^{-1}, write $C = AD$, that is, C is the matrix obtained from

* See references [7, 79].

A by multiplying the jth column of A by t_j. Then

$$C^{-1} = (AD)^{-1} = D^{-1}A^{-1},$$

or

$$A^{-1} = DC^{-1},$$

so that A^{-1} may be found from C^{-1} by multiplying the jth row of C^{-1} by t_j. Note that the elements of the transformed A matrix and of b are all less than 1 in absolute value.

2. Apply the transformation in 1 above to the symmetrized matrix $A^{*\prime}A$. Then all diagonal elements of the resulting matrix are unity, and all non-diagonal elements are between -1 and $+1$. The right-hand side, $A^{*\prime}b$, is also transformed as in 1, and the resulting vector has elements <1 in absolute value.

3. If A is symmetric and positive definite, then a scaling may be performed by multiplying row i and then column i by $1/\sqrt{a_{ii}}$, which makes the resulting diagonal elements equal to 1. The right-hand side $b_i/\sqrt{a_{ii}}$, is normalized by dividing each component by the length of the new b vector.

4. If A is Hermitian (or symmetric) positive definite, then its eigenvalues are all real and positive. Let $\lambda_{\max} \leq \mu$ and form a new system from $Ax = b$ by dividing both A and b by μ. Let $A = A/\mu$, $b = b/\mu$. The new system has real positive eigenvalues between 0 and 1.

5. If A is an arbitrary nonsingular matrix and $Ax = b$, write $A^{*\prime}Ax = A^{*\prime}b$. Then $A^{*\prime}A$ is Hermitian positive definite and has real positive eigenvalues. Let $\lambda_{\max} \leq \mu$ where $\lambda_{\max}$ is the largest eigenvalue of $A^{*\prime}A$. As in 4 above, write $A = (A^{*\prime}A)/\mu$, $b = (A^{*\prime}b)/\mu$ so that the new system has real, positive eigenvalues between 0 and 1.

7

WORK REQUIRED —NUMBER OF ARITHMETIC OPERATIONS [13, 17 19, 33, 38]

The number of divisions, multiplications, additions, and recordings of intermediate results provides a very rough estimate of the efficiency of an algorithm. For each direct method and for each step of an iterative method it is possible to estimate the numbers of these operations as functions of n, the order of the matrix. Such functions, however, may well be discontinuous functions of n, that is, of one form when all initial and computed quantities fit into the internal memory of the machine, but of quite another form (possibly even an entirely different method) when n is so large that auxiliary storage is required, involving the transfer of information between internal and external memories.

Thus the "number of operations required" as stated in the following table as a function of n, is given with some trepidation. Such information should be used cautiously and should not be used as even a rough measure of the *time* required without careful analysis. Is the method such that a large number of zeros in the matrix reduces the work? How many iterations are needed for the desired level of convergence? Is the computer such that several arithmetic operations may be performed concurrently (as, e.g., the CDC 6600)? Do branching and decision operations require more or less time on the computer than certain arithmetic operations?

In Table 7.1 below those numbers that stretch into both $\div$ and $\times$ columns indicate that the number is the lumped sum of multiplications and divisions needed. Not all methods are included in the table. Boldface items indicate preliminary operations which are performed once, but not repeated for each iteration.

TABLE 7.1
Number of arithmetic operations
(A is an $n \times n$ matrix)

Method	$AX = b$			A^{-1}		
Direct	$\div$	$\times$	$+$	$\div$	$\times$	$+$
Gaussian Elimination	n	$\frac{1}{3}n^3 + n^2 - \frac{1}{3}n$	$\frac{1}{3}n^3 + n^2 - \frac{5}{6}n$	n	$n^3 - 1$	$n^3 - 2n^2 + n$
Jordan	n	$\frac{1}{2}n^3 + n^2 - \frac{1}{2}n$	$\frac{1}{2}n^3 - \frac{1}{2}n$	n	$n^3 - 1$	$n^3 - 2n^2 + n$
Doolittle	n	$\frac{1}{3}n^3 + n^2 - \frac{1}{3}n$	$\frac{1}{3}n^3 + n^2 - \frac{5}{6}n$	n	$n^3 - n$	$n^3 - 2n^2 + n$
Cholesky (symmetric)	n root recip.	$\frac{1}{6}n^3 + \frac{3}{2}n^2 + \frac{1}{3}n$	$\frac{1}{6}n^3 + n^2 - \frac{7}{6}n$	n	$\frac{1}{2}n^3 + \frac{3}{2}n^2 - n$	for $(L')^{-1}L^{-1}$
					$\frac{1}{2}n^3 + n^2 - \frac{1}{2}n$	for $(L'A^{-1}) = L^{-1}$
Crout		$\frac{1}{3}n^2 + \frac{1}{2}n^2 + \frac{1}{6}n$				
Triangular system	n	$\frac{1}{2}n^2 - \frac{1}{2}n$	$\frac{1}{2}n^2 - \frac{1}{2}n$	n	$\frac{1}{6}n^3 + \frac{1}{2}n^2 - \frac{2}{3}n$	$\frac{1}{6}n^3 - \frac{1}{2}n^2 + \frac{1}{3}n$
Product form of the inverse		$\frac{1}{2}n^3 + \frac{3}{2}n^2$			$\frac{1}{2}n^3 + \frac{1}{2}n^2$	
Tridiagonal	$2n$	$4(n - 1)$	$4(n - 1)$			
Orthogonal		$2n^3 + 2n^2$				
Aitken: $CA^{-1}B$ C is $p \times n$; B is $n \times m$		$\frac{1}{3}n^3 + \frac{1}{2}(m + p)n^2 + (mp - \frac{1}{3} - \frac{1}{2}m + \frac{1}{2}p)n$				
Rank annihilation		—			$\frac{7}{3}n^3$	

Iterative	Per Iteration			
	÷	×	+	
Jacobi and Seidel	**n**	n^2	$n^2 - n$	
Steepest descent (gradient)	1	$2n^2 + 3n$	$2n^2 + 2n - 2$	
Conjugate gradient (not symm.)	1	$3n^2 + 6n + 2$	$3n^2 + 3n - 3$	
Monte-Carlo	**n**	$\mathbf{n^2}$	$\mathbf{n^2 + 2n}$	
	1	$n(W + 2)$	$n[(n - 1)W + 3]$	W = average number of steps per random walk
Successive overrelaxation	**n**	$n(n + 1)$	n^2	
Peaceman–Rachford			**5n**	
	n	$2n^2 + 8n - 8$	$2n^2 + 8n - 8$	$A = A_1 + A_2 + D$, A_1 and A_2 *tridiagonal*
Newton–Raphson Analogue	0	$2n^3$	$2n^3 - 2n^2 + n$	

REMARKS ON THE TABLE OF ARITHMETIC OPERATIONS (TABLE 7.1)

1. When the matrix is symmetric, the number of operations is reduced by about half.

2. For some of the methods, several different counts could be given, corresponding to slightly different forms of the equations. Some of these counts are only approximate—some authors drop out the lower-order powers of n.

3. For all of the triangularization methods, the number of multiplications required for solving $Ax = b$ is about the same, $\frac{1}{3}n^3$. For the Jordan method this number is about $\frac{1}{2}n^3$. Hence Gaussian elimination is more efficient than the Jordan method for solving systems of linear equations. There is no difference when finding A^{-1}. The product form of the inverse requires the fewest operations for finding A^{-1}.

4. Orthogonal methods and rank annihilation appear to require more work than the other direct methods given in the table. Cramer's rule and the method based upon the Cayley–Hamilton theorem (not given in the table) are even more laborious.

5. An iterative method applied to a full matrix would need to converge in less than n steps to bring its operations count down to that of a direct method.

6. From the table it is clear that A^{-1} should never be found merely for the purpose of finding $A^{-1}B$, since solving $AX = B$ directly is far more efficient. In the case of Gaussian elimination, for example, solving $AX = B$ directly takes about $(\frac{1}{3}n^3 + kn^2 - \frac{1}{3}n)$ multiplications, whereas finding X by first computing A^{-1}, then $A^{-1}B$ takes about $(n^3 + kn^2 - 1)$ multiplications, where k is the number of columns in B.

8

COMMENTS AND COMPARISONS

The following factors should be considered when comparing numerical methods for solving systems of linear equations or for inverting matrices:

1. Accuracy obtained—stability with respect to roundoff.
2. Time required—convergence rate for iterative methods.
3. Applicability—type of matrix to which the method applies.
4. Each step of the process should be an improvement in the approximation.
5. As much of the original data as possible should be used.
6. Storage requirements.
7. Ease of coding.
8. Performance on ill-conditioned systems.

8.1 COMMENTS ON DIRECT METHODS

1. Direct methods find a matrix inverse or solve a system of linear equations using a finite number of operations which can be specified in advance, so that the number of operations performed is independent of the accuracy desired. An exact solution would be obtained if there were no roundoff errors. The accuracy of the approximate solution obtained by a digital computer is a function of the condition and size of the matrix, the accuracy of the arithmetic as performed by the computer, and the algorithm used for solution.

2. The original matrix is altered as the computations proceed and there is no particular advantage to be gained when the matrix has a large number of

zeros systematically arranged, except, of course, in the case where the algorithm is designed specifically for this matrix form, for example, for a tridiagonal matrix.

3. Most of the direct methods apply to quite general matrices, real or complex, and are well adapted to either matrix inversion or solution of systems of linear equations with one or many right-hand sides.

4. Storage requirements are about the same for most of the general direct methods, a minimum of about $n^2 + rn$, where r is the number of right-hand sides. Storage for the program itself and for keeping track of any row or column permutations is usually small compared with that used by the matrices. The method of rank annihilation for finding the inverse matrix requires relatively large storage, about $2n^2 + 4n + 300$, compared with most of the other direct methods. Storage requirements will also depend upon how many of the quantities A^{-1}, $A^{-1}B$, A the user wants available at the same time.

5. The largest general matrix that can be inverted on a given computer is determined by the accuracy required, the condition of the matrix, the storage available, and the time one is willing to spend. Some indications of accuracies obtainable are given in Section 5.2, but these error expressions are functions of a condition number $P(A)$ or of $\|A^{-1}\|$, which may be unavailable. In most cases these error bounds tend to be severely large. Double-precision arithmetic could, of course, increase the accuracy obtainable, but this also increases the storage needed. A variation of Gaussian elimination (see [13], pp. 186–188) for matrices too large to store in high-speed memory may be used to reduce the storage needed to a little more than half that required for the usual method. The method of pivoting is slightly different from partial positioning.

8.2 COMPARISON OF DIRECT METHODS [11, 13, 17, 42, 81, 84]

8.2.1 Real Symmetric Positive-Definite Matrices

The Cholesky square-root method is probably the most efficient. No pivoting is needed. The only adverse comment is that $(n - 1)$ square roots must be computed. If this is particularly troublesome or time consuming, the method of congruent transformations could be used. Since real symmetric positive-definite matrices occur quite frequently in practice, it is probably worth having a special routine available for these matrices.

8.2.2 Real Symmetric Matrices

The symmetric decomposition of a symmetric matrix which is not positive definite has none of the stability of the positive-definite case. Hence pivoting and interchanges must be used and these usually destroy symmetry. Selecting the largest diagonal element at each stage as pivot does preserve symmetry, but does not guarantee stability. The case of the real symmetric matrix then may just as well be treated under the real matrix case given next.

8.2.3 Real Matrices

If both matrix inverse and solution of systems of linear equations are to be found, then Gaussian elimination with either partial or complete positioning is the method most often recommended. Wilkinson [42] prefers partial positioning, especially if accumulation of inner products is available on the computer, and is of the opinion that extensive pivotal growth is remote. The Gauss–Jordan method is considerably slower than Gaussian elimination for solving systems of linear equations, though just as efficient for inverse computation. One of the compact elimination schemes with interchanges for positioning would be superior on a computer with accurate scalar-product accumulation. Compact elimination would also be called for in the case where the *LU* decomposition was desired explicitly. It would appear to be of no advantage to use any of the orthogonalization methods, Cramer's rule, the method based on the Cayley–Hamilton theorem, below-the-line methods, or rank annihilation unless some intermediate computation performed in one of these methods is expressly wanted.

Partitioning methods could be useful if inverses of intermediate matrices are also wanted, or if the matrix to be inverted is too large for storage in the computer's high-speed memory. Partitioning does *not* save time, as is shown by the table in Appendix D.

The algorithms for triangular and for tridiagonal matrices save time over the general Gaussian elimination procedure and make useful subroutines.

The product form of the inverse should be considered in linear programming applications.

8.2.4 Arbitrary Complex Matrices

Gaussian elimination with either partial or complete positioning and with all operations performed in complex arithmetic is probably the most efficient method for complex-matrix inversion and solution of systems of complex linear equations. It is certainly more efficient than partitioning the matrix and solving in terms of real matrices and real operations as discussed in Section 2.6.5. Most of the direct methods may be applied to complex matrices simply by using complex arithmetic instead of real arithmetic. Exceptions are those methods developed specifically for real symmetric positive-definite matrices. These can usually be modified slightly to apply to

Hermitian positive-definite matrices. Methods which use orthogonal matrices may be generalized to use unitary matrices. Modifications usually entail little more than replacing the transpose by the conjugate transpose.

8.3 TESTS OF DIRECT METHODS

8.3.1 Lietzke, Stoughton, and Lietzke [81]

The Jordan, Cholesky, congruent-transformation, and rank-annihilation methods were tested on symmetric positive-definite matrices of orders 10 to 30. The matrices used were test matrices $-A_{15}$, ${A_{15}}^2$, $-{A_{15}}^3$, A_5, and A_6 of Appendix C. Their conclusions were that any of the methods tested would invert a well-conditioned matrix satisfactorily, but that when the matrix is ill-conditioned, the Jordan method appears superior. If memory capacity is a problem so that a triangular array is desirable, then the Cholesky and congruent-transformation methods offer an advantage for large n. Only for $-{A_{15}}^3$ was Cholesky the better of the two. Rank annihilation was not particularly successful because of some peculiarity of the program.

8.3.2 Author's Tests

The direct methods tested on the CDC 1604A computer were Gauss–Jordan elimination with both partial and complete positioning, Cholesky method, Crout method with partial pivoting, rank annihilation, and partitioning. Inverses were found for each of the test matrices $A2$, $A4$, $A5$, $A6$, $A7$, $A8$, $A10$, $A13$, $A14$, $-A15$, $(A15)^2$, and $A19$ for $t = 0.05$, all from Appendix C. Condition numbers $N(A)$ and $M(A)$, (see Chapter 4), error measures f_r and f_e, (see Chapter 5), and computation time in seconds were recorded for each case.

In general the method of rank annihilation proved much slower (except in one case), no more accurate and less often applicable than the other four direct methods. Accuracies of these other direct methods were about the same except for $-A15$ where Gauss–Jordan with complete pivoting was superior. The Cholesky method, which was designed for symmetric positive-definite matrices, was fastest when applicable. Next in order of speed were Gauss–Jordan with partial pivoting, Crout with partial pivoting and Gauss–Jordan with complete pivoting. For $n = 50$, average computing times were 9.7, 15.2, 16.6, and 20.1 sec respectively. It is interesting to compare the actual computing time with that predicted by using the table of arithmetic operations required (in Chapter 7) together with the addition, multiplication, and division speeds of the computer.

TABLE 8.1
Computing speeds for CDC 1604A and CDC 6600

1604A	6600	
$56(10^{-6})$ sec	$2.9(10^{-6})$ sec	Divide
$45(10^{-6})$ sec	$1.0(10^{-6})$ sec	Multiply
$19(10^{-6})$ sec	$0.4(10^{-6})$ sec	Add

Consider only Gaussian (or Gauss–Jordan) elimination for finding A^{-1}. The number of operations, which is the same for both methods, is

$$n \quad \text{divisions,}$$

$$n^3 - 1 \quad \text{multiplications,}$$

and

$$n^3 - 2n^2 + n \quad \text{additions.}$$

Computing speeds for the CDC 1604A and CDC 6600 computers are given in Table 8.1.
Predicted computing time for the CDC 1604A is then

$$10^{-6}(64n^3 - 38n^2 + 75n - 45) \text{ seconds,}$$

and for the CDC 6600 is

$$10^{-6}(1.4n^3 - 0.8n^2 + 3.3n - 1) \text{ seconds.}$$

For $n = 50$, the actual and predicted computing times in seconds are given in Table 8.2.
(The actual "figures given for the CDC 6600 were computed from the approximation formula, $2.3n^3$ to $2.5n^3$ μsec, which was derived empirically, from many runs on matrices of various sizes, by K. Ellenberger.) Hence there is a "fudge factor" by which each prediction formula must be multiplied that accounts for operations other than the basic arithmetic ones and that depends on the computer used and on the particular program used. This factor can only be determined by running the program many times on the same computer with varying types and sizes of matrices.

TABLE 8.2
Computing times (seconds)

	1604A	6600
Predicted	7.908705	0.173164
Actual	16.17 to 16.63	0.2875 to 0.3125

8.4 COMMENTS ON ITERATIVE METHODS [5, 11–13, 40, 66, 75, 108–110]

1. In general iterative methods are preferred for solving a single large sparse system of linear equations for which convergence is known to be rapid. Iterative methods are less useful for finding inverses or for solving systems with many right-hand sides. Iterative procedures begin with an approximate solution to a linear system and obtain an improved solution with each step of the process. An iterative process would require an infinite number of steps to obtain an exact solution; the accuracy of the solution obtained depends upon the number of iterations performed (and/or the convergence rate), the condition of the matrix as well as its size, the accuracy of the arithmetic performed by the computer and the particular algorithm used.

2. It is purported that iterative methods use the matrix in its original form for each iteration, hence tend to be self-correcting and to minimize roundoff error. Wilkinson maintains that the idea that one works with the original matrix is an illusion and challenges the statement that iterative methods are less affected by rounding errors. Iterative methods do have the advantage of preserving the zero elements of the matrix.

3. Iterative methods are particularly efficient for solving systems of linear equations arising from finite-difference approximations for elliptic partial differential equations. The associated matrices in such cases are characterized by a great many zero elements and by the fact that the nonzero elements occur in some systematic pattern. In addition, such matrices are usually symmetric positive definite and irreducible. For very large-order matrices, direct methods would be too laborious, require too much storage, and give limited accuracies. Iterative methods, on the other hand, are ideally suited to such matrices, particularly when the pattern of nonzeros is a cyclic one needing little storage even for very large-order matrices. For example, Varga [40] states that the Bettis Atomic Power Lab of Westinghouse Electric had in daily use in 1960 a two-dimensional program which would treat as a special Laplacian-type matrix equations of order 20,000 by cyclic iteration.

4. The total amount of work involved in using an iterative method depends upon the convergence rate and the desired accuracy. Usually an iteration is used only in problems for which convergence is known to be rapid. Slow or irregular convergence would be a considerable drawback. An iteration for which convergence is assured in say less than n steps would be quite good. Since one step of an iterative method involves roughly n^2 multiplications (see Chapter 7), the number of iterations needed for sufficient accuracy would need to be less than $\frac{1}{3}n$ to beat an elimination method. The rate of convergence

for an ill-conditioned matrix will be poor. In very ill-conditioned systems where the largest eigenvalue of the iteration matrix is near 1, the change between successive iterates is so slight that no progress in convergence is made. (If $|A| \sim 0$, then $\lambda = 1$ is nearly an eigenvalue of the iteration matrix.) Many iterative schemes are dependent upon the ordering of the equations, and the rate of convergence may be improved by suitable row and column permutations.

5. The storage required for an iterative method will vary with the matrix used. In the worst case where all of A, $x^{(k)}$, $x^{(k+1)}$, and b are stored to solve

TABLE 8.3

Storage requirements of iterative methods for solving $Ax = b$, where A is an $n \times n$ matrix and x and b are n-component vectors

Method	Approximate Storage Needed
Jacobi	$n^2 + 3n$
Seidel	$n^2 + 2n$
Back-and-forth Seidel	$n^2 + 2n$
SOR(Extrapolated Seidel)	$n^2 + 2n$
Richardson	$n^2 + 3n + 2k$[a]
Extrapolated Jacobi	$n^2 + 3n$
Peaceman–Rachford	$n^2 + 3n$ or $(15n - 8)$[b]
Monte-Carlo	$n^2 + 4n + 2$
Gradient (steepest descent)	$n^2 + 4n + 2$
Conjugate gradient (nonsymmetric)	$n^2 + 6n + 3$
Newton–Raphson analogue	$4n^2$

[a] Where k is the chosen number of parameters.
[b] When $A = A_1 + A_2 + D$ and A_1 and A_2 are tridiagonal.

$Ax = b$, a total of $n^2 + 3n + \alpha$ cells are needed, where α is the number needed for program instructions. In most cases where iteration is to be used only a small proportion of the n^2 elements of A require storage. In the best cases there may be as few as three nonzero elements per row and if these nonzero elements can be generated by a simple formula rather than stored, storage is indeed minimal. Approximate storage requirements of various iterative methods for solving $Ax = b$ are given in Table 8.3.

8.5 COMPARISON OF ITERATIVE METHODS [50, 110, 113]

8.5.1 Jacobi; Seidel

The Jacobi method (and/or the Seidel method) serves as a yardstick against which most of the other iterative methods may be measured. Although the

Seidel method converges twice as fast as the Jacobi method, neither is usually considered efficient enough for use with the large sparse matrices (derived from partial differential equations) for which iterative methods are appropriate.

8.5.2 Conjugate Gradients, Lanczos, and Gradient

The method of conjugate gradients has not been widely adopted for solution of these large linear systems because of the relatively high storage requirements and relatively complex structure of each iterative step. The same is true for the Lanczos and gradient (steepest-descent) methods. In the latter case, convergence is only linear and not very effective without an accelerating procedure.

8.5.3 Newton–Raphson Analogue

The Newton–Raphson analogue requires too many operations and too much storage even though convergence, when it occurs, is of second order.

8.5.4 Extrapolated Jacobi; Extrapolated Seidel

Extrapolated Jacobi is equivalent to Richardson's method with a single parameter; extrapolated Seidel is equivalent to successive overrelaxation, so these methods need not be considered separately.

8.5.5 Monte-Carlo, Stationary Linear, and Gaussian Elimination

Section 3.3.1.4 gives a comparative table for the number of multiplications necessary to find *one* component of the solution vector of a linear system, to a given accuracy, for the Monte-Carlo, stationary linear iteration, and Gaussian-elimination methods. The table indicates that a Monte-Carlo method might be useful in a case where only a rough approximation is needed, or in the case where n is so large that the other two methods are impossible or very inefficient.

8.5.6 SOR, Peachman–Rachford, Douglas–Rachford, and Richardson

The iterative methods most frequently used and which have proved most efficient in practice are SOR and its variations, Peaceman–Rachford, and, to a lesser extent, Douglas–Rachford and Richardson's methods. Some insight into the comparative behavior of these methods may be gained from the remarks below as well as from the reports which follow giving results of test studies conducted by various researchers.

1. SOR is faster, needs less storage and causes less roundoff error than Richardson's method [110].

2. For the Helmholtz partial differential equation in the unit square the optimized one-parameter Peaceman–Rachford method and optimized *point* SOR methods have the same asymptotic rates of convergence for all mesh widths greater than zero. However, m-parameter Peaceman–Rachford methods are superior in convergence rate for $m > 1$ to *point* SOR methods for all sufficiently small mesh spacings [50].

3. In general the number of iterations needed to achieve a specified accuracy varies as $|\log \delta|^{-1}$ for the Peaceman–Rachford method, which is more favorable than the δ^{-1} necessary for SOR as the mesh width δ decreases in size [113].

4. SOR is simpler than the Peaceman–Rachford method and has been extended to a wide class of partial differential equations over nonrectangular regions. The Peaceman–Rachford formulas are more complex and require that data be obtained by "columns" then by "rows." The theory for the Peaceman–Rachford method is limited to a restricted class of partial differential equations over a rectangular region, although the method has been applied to other cases successfully [113].

5. The Douglas–Rachford method is inferior to Peaceman–Rachford with either the Peaceman–Rachford or Wachspress parameters, though it is convergent for either choice [50].

6. Block SOR is preferable to point SOR whenever the sufficient conditions of Section 3.3.2.3 are satisfied.

8.6 TESTS OF ITERATIVE METHODS

8.6.1 Birkhoff, Varga, and Young [50]

1. Numerical experiments were performed approximating Laplace's equation, $\partial^2 x/\partial u^2 + \partial^2 x/\partial v^2 = 0$, with a five-point difference equation over five different regions, and assuming all boundary values zero. An iteration was stopped when values at all mesh points were less than 10^{-6} in absolute value. The Peaceman–Rachford method was tested with three sets of parameters—Peaceman–Rachford parameters, Wachspress parameters, and optimum parameters. The number, k, of iteration parameters used ranged from 1 to 5. For point SOR the optimum relaxation factors were determined and used. In the square region where the number of iterations for the Peaceman–Rachford method could be predicted, the observed number of iterations agreed closely with the predicted number.

The Peaceman–Rachford method proved very effective, and for small δ and $k > 1$, was much superior to the point SOR. The number of iterations for the Peaceman–Rachford process was bounded by $\delta^{-1/k}$, for fixed k, and by $|\log \delta|$ for a "good" k value. Hence the increase in computer time involved in passing to a smaller mesh size is almost entirely due to the increase in the number of points and only very slightly due to an increase in the number of iterations.

Wachspress iteration parameters are recommended in preference to Peaceman–Rachford or optimum parameters (the latter are difficult to determine in general) for the Peaceman–Rachford method. It is wiser to choose a value of k, the number of iteration parameters, that is too large than to use one that is too small, and it is recommended that, in the absence of other information, the number of parameters be chosen between $1\frac{1}{2}$ and 2 times the k value obtained by the procedure of Section 3.3.2.2 (a).

Although the number of iterations for SOR was slightly larger than for the Peaceman–Rachford method with $k = 1$, only half as much time was required per iteration with SOR; hence SOR was definitely superior to Peaceman–Rachford with *one* parameter. Since the number of iterations with SOR is asymptotically proportional to $(2\pi\delta)^{-1}$ as compared to $\frac{1}{4}k(2/\pi\delta)^{1/k}$ for the Peaceman–Rachford with Peaceman–Rachford parameters, the latter is superior for $k > 1$, which was borne out by the numerical results in all regions. For example, when $\delta = 1/160$, the number of iterations for SOR was about 570 as compared with 22 iterations using the Peaceman–Rachford method with $k = 5$ Wachspress parameters.

2. Using the same problem and regions as in 1 above, experiments were performed comparing the SOR variants with Peaceman–Rachford methods. Arithmetic requirements per mesh point of the multiline (K-line) SOR methods increase linearly with K. Increasing K in the SOR methods yields an improved asymptotic rate of convergence bounded by δ as $\delta \to 0$. Peaceman–Rachford methods have asymptotic convergence rates bounded by $\delta^{1/k}$ for this problem.

A two-line (block) SOR coupled with a cyclic Chebyshev acceleration was compared with the Peaceman–Rachford iteration. For every δ each process was optimized with respect to acceleration parameters. There was considerable decrease in iteration time in passing from point SOR to two-line cyclic Chebyshev SOR. It was found that a critical value, δ^*, exists so that if $\delta > \delta^*$, it is better to use two-line cyclic Chebyshev, but for all $\delta < \delta^*$ the optimized (the number of parameters to be used cyclically was varied to obtain the optimum) Peaceman–Rachford is superior in terms of actual machine time. The Peaceman–Rachford for very small δ was vastly superior to any of the SOR variants. The critical value δ^* of δ varies greatly from problem to problem.

8.6.2 Sheldon [93]

Studies were conducted to determine the number N of iterations necessary to reduce the Euclidean norm of the error vector by a factor of 10^{-3} for the

Dirichlet problem in a square with 50 equal mesh spacings. The following results were obtained:

For unaccelerated Jacobi:	$N \leq 3500$
For Chebyshev accelerated Jacobi:	$N \leq 120$
For SOR with $\omega = \omega_{\text{opt}}$:	$N \leq 81$
For modified Chebyshev accelerated SOR with $\omega = 1$:	$N \leq 63$.

8.6.3 Forsythe and Wasow [12]

Convergence rates were found using Poisson's equation with $x(u, v)$ fixed on the boundary. A square region of side δ was assumed and divided into n^2 squares with $n\delta = \pi$ giving $(n - 1)^2$ interior points. In some cases both five- and nine-point finite difference approximations were used for the partial differential equation. Approximate convergence rates are listed for each of several methods in Table 8.4.

TABLE 8.4
Convergence rates

Method	Approximate Rate of Convergence
Jacobi	$\frac{1}{2}\delta^2$
Seidel	δ^2
Richardson (k parameters)	
$k \sim \infty$	δ
$k \sim 2n$	0.89δ
$k \sim n$	0.78δ
$k \sim \frac{1}{2}n$	0.59δ
Optimum point SOR	
5-point difference equation	2.00δ
9-point difference equation	2.04δ
Block Jacobi (two-line)	δ^2
Block Seidel (two-line)	$2\delta^2$
Optimum block SOR (two-line)	$2\sqrt{2}\delta$
Peaceman–Rachford	$>0.777/\ln(1/\delta)$
Douglas–Rachford	$>0.535/\ln(1/\delta)$

8.6.4 Young [111]

For the Dirichlet problem on the unit square, mesh width δ, estimates of the convergence rates and number of iterations necessary to reduce the error to 0.001 of its initial value were obtained for the SOR method with $\omega = 1$ and with $\omega = \omega_{\text{opt}}$. The results are summarized in Table 8.5.

TABLE 8.5
Estimates of rates of convergence and number of iterations for reducing error to 10^{-3} of its original value

δ^{-1}	Gauss–Seidel, $\omega = 1$ Rate	Iterations	ω_{opt}	SOR Rate	Iterations
20	0.024776	279	1.729454	0.315459	35
50	0.003950	1749	1.881839	0.125746	92
100	0.000998	6922	1.939091	0.062843	195
300	0.000110	62798	1.979272	0.020946	640

8.6.5 Author's Tests

Limited tests for iterative methods of solving the system $Ax = b$ (where b was the vector of all 1's) were performed on the CDC 1604A computer. The matrices A used were those from Appendix C as listed under tests of direct methods. Error measures f_r and f_e, the total time in seconds and the number of iterations were recorded for each test. The iterative methods tested were Seidel, back-and-forth Seidel, conjugate-gradient for symmetric positive-definite matrices, and conjugate-gradient for arbitrary matrices.

Neither of the Seidel methods proved very satisfactory since convergence was slow and acceleration procedures were not used. Little difference in accuracy appeared between Seidel and back-and-forth Seidel.

The conjugate-gradient methods were superior to the two Seidel methods. Compared with direct methods such as Gauss–Jordan elimination and Cholesky, the gradient methods show similar accuracies but far greater variability in the time required for solution. For example, in the conjugate-gradient symmetric positive-definite case for $n = 50$, the time required to solve $Ax = b$ ranged from 0.367 to 33.58 sec, whereas for Gauss–Jordan with partial pivoting and $n = 50$, the time needed to solve $AX = I$ varied from 12.35 to 16.63 sec. Also for $n = 50$, the Cholesky positive-definite method took 7.93 to 10.67 sec to solve $AX = I$.

Computer time was not available for further study which would have included acceleration procedures, Richardson's method, the SOR, and ADI methods, and some of the block SOR procedures, all of which have proved successful in solving the special systems arising from elliptic partial differential equations.

8.7 SUMMARY OF THE CONVERGENCE PROPERTIES OF THE JACOBI, SEIDEL, SOR, AND RICHARDSON ITERATIVE METHODS [109]

Properties of Matrix A

1. $a_{ii} > 0$
2. Symmetric

3. Positive definite
4. $a_{ij} \leq 0$ for $i \neq j$
5. $\sum_{\substack{j=1 \\ j \neq 1}}^{n} |a_{ij}| \leq a_{ii}$ and for some i the "less than" sign holds
6. Irreducible
7. Property A

Note that 1, 2, 5, and 6 imply 3.

Notation

1. $A = D + L + U$ where $D = (a_{ii})$, $U = (a_{ij})$, $i < j$, and $L = (a_{ij})$, $i > j$.
2. General linear iteration: $x^{(k+1)} = T^{(k)}x^{(k)} + q^{(k)}$
3. Iteration matrices:

   Stationary case: $T^{(k)} = T$ for all k

   Nonstationary case: $M_k = \prod_{i=0}^{k-1} T^{(i)}$

Spectral radius of T or of M_k is denoted $\lambda(T)$ or $\lambda(M_k)$
Asymptotic rate of convergence is $R(T) = -\log \lambda(T)$
Average rate of convergence for given k is $R_k = R(M_k) = -\log \lambda(M_k)/k$

Method	Iteration Matrix	Spectral Radius	Rate of Convergence	Convergence Occurs if A Satisfies Properties
Jacobi	$T_J = I - D^{-1}A$	$\lambda = \lambda(T_J)$	$R(T^J) = -\log \lambda$	1–4 1–3, 7 1, 5, 6 1, 4; Seidel converges 1, 7; Seidel converges
Seidel	$T_1 = I - (D + L)^{-1}A$	$\mu = \lambda(T_1) = \lambda^2$	$R(T_1) = 2R(T_J)$	1–3 1, 5, 6 1, 4; Jacobi converges 1, 7; Jacobi converges
SOR	$T_\omega = I - (\omega^{-1}D + L)^{-1}A$ where $0 < \omega < 2$			
	$\omega_{\text{opt}} = \dfrac{2}{1 + (1 - \lambda^2)^{1/2}}$	$(\lambda T\omega_{\text{opt}}) = \omega_{\text{opt}} - 1$	$R(T\omega_{\text{opt}}) = 2\sqrt{2}(R_J)^{1/2}$	1–3, 7; $0 < \omega < 2$.
			$\sqrt{R(T_1)} \leq R(T_\omega) \leq 2\sqrt{R(T_1)}$	1–3, 6; $0 < \omega < 2$. 1–4; $0 < \omega < 2$. 1–3, 7; scalar diagonal → SOR converges more than twice as fast as best Richardson

Method	Iteration Matrix	Spectral Radius	Rate of Convergence	Convergence Occurs if A Satisfies Properties
Richardson	$T^{(k)} = (I - \alpha_k A)$			1, 2; $k = 1$, $0 < \alpha_0 < 2/b'$.
	a' = min eigenvalue of A			1–3; $0 < \alpha_j < 2/b'$.
	b' = max eigenvalue of A			1–3, 7; $k = 1$, $\alpha_0 = -1/a_{ii}$, scalar diagonal $\rightarrow$ Richardson = Jacobi.
	$y_j^{(k)} = \cos\left[\frac{(2j+1)\pi}{2k}\right]$	$\lambda(M_k) = \frac{1}{T_k(y_0)}$	$R(M_k)$	1–3; α_j at left
	$j = 0, \ldots, k-1.$	$T_k(y_0)$	$= \frac{1}{k} \log T_k(y_0)$	
	$\alpha_j = 2[(b' + a') - y_j^{(k)}(b' - a')]^{-1}$	$= \frac{1}{2}[q^k + q^{-k}]$		
	$y_0 = \frac{(b' + a')}{(b' - a')}$			
	$q = y_0 + (y_0^2 - 1)^{1/2}$			

APPENDIX A

GLOSSARY OF MATRIX TERMINOLOGY

In general, $A = (a_{ij})$ will be an arbitrary matrix of n rows and n columns. An attempt has been made to include most of the terms used in the text and concerning matrix algebra. Some of these may also be defined again in the text.

Adjoint, (A_{ji}), of $A = (a_{ij})$ is the transposed matrix of cofactors of A.

A-orthogonal vectors p_i are vectors for which the scalar product $(p_i, Ap_j) = 0$ for $i \neq j$.

Band (codiagonal, striped) matrix is one in which all elements are zero except those appearing on or near the main diagonal, that is, $a_{ij} = 0$ for $|i - j| > m$.

Binomial coefficient, $\binom{N}{R}$:

$$\binom{N}{R} \equiv \frac{N!}{R!\,(N - R)!}$$

Characteristic equation of the matrix A is the polynomial equation, $P(\lambda) = |A - \lambda I| = 0$, where $P(\lambda)$ is a polynomial of degree n in λ and n is the order of A.

Characteristic values or roots (eigenvalues, latent values, principal values, proper values) of the matrix A are the n roots $\lambda_1, \ldots, \lambda_n$ of the characteristic equation of A.

Characteristic vector (eigenvector, latent vector, principal axis, proper vector) corresponding to the characteristic value λ is a vector $x \not\equiv 0$ satisfying the equation $Ax = \lambda x$.

Cofactor, A_{ij}, of a matrix element a_{ij} is $(-1)^{i+j}$ times the determinant of the matrix obtained from A by deleting the ith row and jth column.

Companion matrix of the minimal or of the characteristic polynomial of a matrix A is the matrix C which has 1's just below (above) the main diagonal, the negatives of the polynomial coefficients in the last column (row), and zeros elsewhere.

Let

$$P(\lambda) = \sum_{i=1}^{n} p_i \lambda^{n-i} + \lambda^n.$$

Then

$$C = \begin{bmatrix} 0 & & & & -p_n \\ 1 & \ddots & & 0 & -p_{n-1} \\ & \ddots & \ddots & & \vdots \\ & & \ddots & 0 & \vdots \\ & 0 & & \ddots & \vdots \\ & & & 1 & -p_1 \end{bmatrix} \quad \text{or}$$

$$C = \begin{bmatrix} 0 & 1 & & & 0 \\ & \ddots & \ddots & & \\ & & \ddots & \ddots & \\ & 0 & & 0 & 1 \\ -p_n & -p_{n-1} & \cdots & & -p_1 \end{bmatrix}.$$

Congruent matrices; A and B are congruent if and only if $A = P'BP$, where P is a nonsingular matrix.

Conjugate $\bar{z}$ of the complex number $z = a + bi$ is $\bar{z} = a - bi$.

Conjugate A^* of the matrix A is the matrix obtained by replacing each element a_{ij} of A by its conjugate $\bar{a}_{ij}$.

Conjunctive matrices: A and B are conjunctive if and only if $A = (P^*)'BP$, where P is a nonsingular matrix.

Deflation: The reduction of the problem of finding the eigenvalues of an $n \times n$ matrix to that of finding the eigenvalues of an $(n-1) \times (n-1)$ matrix after one eigenvalue and eigenvector of the larger matrix have been found. Any technique used after finding one eigenvalue and eigenvector to reduce the original problem to one in which the known value and vector are no longer present.

Determinant of matrix A written $|A|$, d(A), or det(A): The sum of the $n!$ terms $|A| = \Sigma(\pm)a_{1i}a_{2j}a_{3k} \cdots a_{n1}$, where the second subscripts are a permutation of the numbers 1, 2, 3, . . . , n. The even permutations, which contain an even number of inversions, are given a plus sign, whereas the odd ones are made negative.

Diagonal matrix D is one whose only nonzero elements d_i are those along the main diagonal and may be written $D = \text{diag}(d_1, \ldots, d_n)$.

Eigenrow corresponding to the characteristic value λ is a vector $y \not\equiv 0$ satisfying the equation $y'A = \lambda y'$.

Elementary divisors: See Smith form.

Elementary row (column) operations:

1. Interchange rows (columns) r and s of matrix A. This may be accomplished by multiplying A on the left (right) by the matrix $E = (e_{ij})$ defined as $e_{ii} = 1$ except that $e_{rr} = e_{ss} = 0$, $e_{ij} = 0$ for $i \neq j$ except $e_{rs} = e_{sr} = 1$.
2. Multiplications of row (column) r of A by a constant k. This is done by multiplying A on the left (right) by the matrix $E = (e_{ij})$ defined as $e_{ii} = 1$, except $e_{rr} = k$, $e_{ij} = 0$ for $i \neq j$.
3. Addition to row (column) r the kth multiple of row (column) s. Multiply A on the left (right) by the matrix $E = (e_{ij})$ defined as $e_{ii} = 1$, $e_{ij} = 0$ for $i \neq j$ except $e_{rs} = k$.

Equivalent matrices: A and B are equivalent if and only if $A = PBQ$ where P and Q are nonsingular matrices. A may be obtained from B by a finite number of elementary operations on the rows or columns of B.

Error vector, e, for an approximate solution $\tilde{x}$ of the linear equations $Ax = b$: $e = A^{-1}b - \tilde{x}$, that is, the error vector is the correct solution vector minus the approximate solution vector.

Gram–Schmidt orthogonalization process: Given a set of n linearly independent n-component vectors $x_1, \ldots, x_n$, find an orthonormal basis of n-component vectors $y_1, \ldots, y_n$ such that y_i is a linear combination of $x_1, \ldots, x_n$ for $i = 1, \ldots, n$.
Set $y_1 = x_1/\|x_1\|$. If $y_1, \ldots, y_r$ have been constructed, define $c_j = (x_{r+1}, y_j)$ for $j = l, \ldots, r$ and set

$$u = x_{r+1} - \sum_{j=1}^{r} c_j y_j.$$

Then define $y_{r+1} = u/\|u\|$. It can be shown that $(y_i, y_j) = \delta_{jj}$.

Hermitian matrix H: One whose transpose is its conjugate. $(H')^* = H$ or $H' = H^*$.

Hessenberg form: "Almost" triangular form, that is, $a_{ij} = 0$ if $j > i + 1$.

Identity matrix, I A square matrix with 1's along the main diagonal and zeros elsewhere.

$$
\begin{aligned}
I = (\delta_{ij}) &= 0 \quad \text{for} \quad i \neq j; \\
&= 1 \quad \text{for} \quad i = j.
\end{aligned}
$$

Ill-conditioned system of linear equations: In general, a system for which small errors in the coefficients or in the right-hand sides may correspond to very large errors in the solution.

Inverse matrix A^{-1} of A: Assuming that A is nonsingular, A^{-1} is the matrix such that $A \cdot A^{-1} = A^{-1} \cdot A = I$.

Irreducible (Indecomposable) matrix: One that cannot be put in the form $\begin{pmatrix} P & Q \\ 0 & R \end{pmatrix}$, where P and R are square, by simultaneous row and column permutations. A matrix $A = (a_{ij})$ is called irreducible if the following holds: Let S be the set of n numbers $1, \ldots, n$ divided into two disjoint sets $S = T_1 + T_2$. Then for any such subdivision there exists at least one element a_{ij} such that i belongs to T_1 and j to T_2.

Jacobi form of a matrix, A: An upper triangular form with the eigenvalues of A along the main diagonal.

Jordan canonical form, J, of a matrix A:

A quasidiagonal matrix

$$
J = \begin{pmatrix}
J_1 & & & & & & \\
& \cdot & & & & & \\
& & \cdot & & 0 & & \\
& & & J_2 & & & \\
& & & & \cdot & & \\
& 0 & & & & \cdot & \\
& & & & & & J_m
\end{pmatrix}
$$

composed of canonical boxes J_i of the form

$$
J_i = \begin{pmatrix}
J_{i1} & & & & & & \\
& \cdot & & & 0 & & \\
& & \cdot & & & & \\
& & & J_{i2} & & & \\
& & & & \cdot & & \\
& 0 & & & & \cdot & \\
& & & & & & J_{i,m_i}
\end{pmatrix},
$$

where

$$J_{ij} = \begin{pmatrix} \lambda_i & 1 & & & & & 0 \\ & \cdot & \cdot & & & & \\ & & \cdot & \cdot & & & \\ & & & \cdot & 1 & & \\ & & & \lambda_i & \cdot & & \\ & & & & \cdot & \cdot & \\ 0 & & & & \cdot & \cdot & \\ & & & & & \cdot & 1 \\ & & & & & & \lambda_i \end{pmatrix}$$

is of order k_{ij},

$$\sum_{j=1}^{m_i} k_{ij} = k_i,$$

where λ_i is an eigenvalue of A of multiplicity k_i, and $i = 1, \ldots, m$. The number of distinct eigenvalues is m, and

$$\sum_{i=1}^{m} k_i = n$$

is the order of A.

Linear dependence: A set of vectors $\{x_i\}$ is linearly dependent if there exist scalars $\{\alpha_i\}$, not all zero, such that $\Sigma\alpha_i x_i = 0$.

Lower triangular matrix, L: A matrix $L = (l_{ij})$ such that $l_{ij} = 0$ for $i < j$.

Matrix multiplication; If $A = (a_{ik})$, $i = 1, \ldots, r$; $k = 1, \ldots, n$ and $B = (b_{kj})$, $k = 1, \ldots, n$; $j = 1, \ldots, s$ then the matrix $C = AB$ is defined and has elements

$$c_{ij} = \sum_{k=1}^{n} a_{ik} b_{kj}$$

for $i = 1, \ldots, r$; $j = 1, \ldots, s$.

Minimal polynomial of a matrix A is that polynomial $m(\lambda)$ of least degree such that $m(A) = 0$.

Minor of a matrix element, a_{ij}: The (determinant of the) square matrix obtained by deleting the ith row and jth column of A.

Nondefective matrix: One for which all elementary divisors are linear. A matrix which is similar to a diagonal matrix in the complex field.

Nonderogatory matrix: One whose minimal and characteristic polynomials are the same.

Nonnegative definite matrix: A Hermitian matrix whose eigenvalues are all greater than or equal to 0.

Nonsingular matrix: One whose determinant is not zero; hence a matrix whose inverse exists.

Norm (Absolute) of a matrix A **(also called maximum coefficient)** is $\|A\|_A = \max_{i,j} |a_{ij}|$.

Norm (Euclidean) of a matrix A is

$$\|A\|_E = [\text{trace}\,(A^{*\prime}A)]^{1/2} = \left(\sum_{i,j} |a_{ij}|^2\right)^{1/2}.$$

Norm (spectral) of a matrix A is $\|A\|_S = (\text{maximum eigenvalue of } A^{*\prime}A)^{1/2}$.

Norm (of a matrix) subordinate to a vector norm is defined by

$$\|A\| = \max_{x \neq 0} \frac{\|Ax\|}{\|x\|}, \quad \text{or equivalently,} \quad \|A\| = \max_{\|x\|=1} \|Ax\|.$$

Norm (Euclidean) of a vector x is

$$\|x\|_E = (x, x)^{1/2} = \left(\sum_{i=1}^{n} |x_i|^2\right)^{1/2}.$$

Normal matrix A**:** One for which $(A^*)'A = A(A^*)'$.

Order of a square matrix: The number of rows (or columns) of the matrix.

Orthogonal matrix, R**:** A matrix whose transpose is its inverse, that is,

$$RR' = I \quad \text{or} \quad R' = R^{-1}.$$

Orthogonal transformation: $A = R^{-1}BR = R'BR$, where R is an orthogonal matrix. An orthogonal transformation is both a congruence and a similarity.

Orthonormal set of vectors $x_1, \ldots, x_n$ is a set such that the scalar product

$$(x_i, x_j) = \delta_{ij} = 0 \quad \text{for} \quad i \neq j$$
$$= 1 \quad \text{for} \quad i = j,$$

where $i, j = 1, \ldots, n$.

p**-Cyclic matrix:** A matrix is p-cyclic ($p \geq 2$) if, by a permutation of rows and the corresponding columns it can be put in the form below where the D_i are square diagonal matrices and the L_i are rectangular matrices:

$$A = \begin{bmatrix} D_1 & L_1 & 0 & \cdots & 0 & 0 \\ 0 & D_2 & L_2 & \cdots & 0 & 0 \\ \cdot & \cdot & \cdot & \cdots & \cdot & \cdot \\ \cdot & \cdot & \cdot & \cdots & \cdot & \cdot \\ \cdot & \cdot & \cdot & \cdots & \cdot & \cdot \\ 0 & 0 & 0 & \cdots & D_{p-1} & L_{p-1} \\ L_p & 0 & 0 & \cdots & 0 & D_p \end{bmatrix}.$$

Permutation matrix: A matrix with exactly one 1 in each row and column and zeros elsewhere.

Positive-definite matrix:

1. a real symmetric matrix S is positive definite if $x'Sx > 0$ for all vectors $x \not\equiv 0$.
2. A Hermitian matrix H is positive definite if $(x^*)'Hx > 0$ for all vectors $x \not\equiv 0$.

Principal (main) diagonal of a matrix $A = (a_{ij})$ is the diagonal of A whose elements are a_{ii}, $i = 1, \ldots, n$.

Principal minor of a matrix is a minor whose principal-diagonal elements are on the principal diagonal of the matrix.

Principal values of A are the nonnegative square roots of the eigenvalues of $AA^{*\prime}$.

Property A: A matrix $A = (a_{ij})$ has property A if there exist two disjoint subsets S and T of the first n positive integers whose sum is the first n positive integers and such that for any $a_{ij} \neq 0$ either

1. $i = j$, or
2. i is in S and j is in T, or
3. i is in T and j is in S.

This is equivalent to stating that by appropriate permutations of rows and corresponding columns the matrix A can be written in the form

$$A = \begin{pmatrix} D_1 & F \\ G & D_2 \end{pmatrix},$$

where the D_i are square diagonal matrices and F and G are arbitrary rectangular matrices.

Rank of a matrix A is the order of the largest nonsingular matrix which can be formed from A by deleting an equal number of rows and columns.

Rayleigh quotient of a real symmetric matrix A corresponding to the (column) vector x is the scalar defined by $R(x) = (x'Ax)/(x'x)$.

Rayleigh quotient (generalized) of an arbitrary matrix A corresponding to the vectors x (column) and y (row) is the scalar defined by $R(x, y) = (y(\overline{Ax}))/(y\bar{x})$.

Residual vector, r, for an approximate solution $\tilde{x}$ of the linear system $Ax = b$ is $r = b - A\tilde{x}$.

Residual vector, η, corresponding to the eigenvalue λ and eigenvector x of the matrix A is $\eta = Ax - \lambda x$.

Scalar (dot, inner) product of two vectors x and x is

$$(x, y) = x \cdot y = x'y = \sum_{i=1}^{n} x_i y_i,$$

where n is the number of components x_i of x and y_i of y. If x and y have complex components, then define

$$(x, y) = \sum_{i=1}^{n} x_i \bar{y}_i.$$

Similarity invariants: See Smith form.

Similar matrices: A and B are similar if and only if $A = P^{-1}BP$, where P is a nonsingular matrix.

Skew Hermitian: A is skew Hermitian if $A = -(A^*)'$, that is, $a_{ij} = -a_{ji}$.

Skew symmetric: A is skew symmetric if $A = -A'$, that is, $a_{ij} = -a_{ji}$.

Smith form: Let A be a matrix of rank r. Reduce the matrix $(\lambda I - A)$ to diagonal form by an equivalence transformation to obtain

$$P(\lambda I - A)Q = \text{diag}(f_1(\lambda), \ldots, f_r(\lambda)) = D(\lambda),$$

where $f_j(\lambda)$ divides $f_{j+1}(\lambda)$, $j = 1, \ldots, r - 1$ and where $f_j(\lambda)$ has leading coefficient 1. P and Q are products of elementary polynomial matrices. $D(\lambda)$ is called the **Smith canonical form** of A; $f_j(\lambda)$ are the **similarity invariants** of A; and $f_r(\lambda)$ is the **minimal polynomial** of A. The **elementary divisors** of A are the factors of the $f_j(\lambda)$ into irreducible polynomials with leading coefficient 1 and with coefficients which are real or complex according as A is real or complex.

Spectral radius, $\lambda(A)$, of a matrix A is the maximum of the absolute values of the eigenvalues of A.

Symmetric matrix, S is one which is equal to its transpose, that is $S' = S$ or $s_{ij} = s_{ji}$.

Trace (spur) of a matrix, A is the sum of the elements on the main diagonal

$$\sum_{i=1}^{n} a_{ii}.$$

Transpose A' of a matrix A is the matrix obtained from A by interchanging rows and columns, that is, $A' = (a'_{ij}) = (a_{ji})$.

Tridiagonal (Triple diagonal) matrix is one for which the only nonzero elements are along the main diagonal and along the diagonals just above and just below the main diagonal, that is, $a_{ij} = 0$ for $|i - j| > 1$.

Unitary matrix V: One whose transposed conjugate is its inverse, $(V^*)' = V^{-1}$ or $(V^*)'V = I$.

Unitary transformation: $A = V^{-1}BV = (V^*)'BV$ where V is a unitary matrix. A unitary transformation is both a conjunction and a similarity.

Unit upper (lower) triangular matrix: An upper (lower) triangular matrix with 1's along the main diagonal.

Unit vector, ϵ_i, is an n-component vector which has 1 as its ith component and zero for all other components.

Upper left principal minor of A: A submatrix of A of the form

$$\begin{pmatrix} a_{11} & \cdots & a_{1k} \\ \cdot & \cdots & \cdot \\ \cdot & \cdots & \cdot \\ \cdot & \cdots & \cdot \\ a_{k1} & \cdots & a_{kk} \end{pmatrix}.$$

Upper triangular matrix, U: A matrix $U = (u_{ij})$ such that $u_{ij} = 0$ for $i > j$.

APPENDIX B

THEOREMS ON MATRIX ALGEBRA

Proofs and further information may be found in one or more of the following references: [1, 6, 7, 10, 12, 13, 15, 19, 25, 27, 28, 32, 38, 39, 41, 60].

MATRIX DECOMPOSITION

1. *LDU* **Theorem.** Provided the upper-left principal submatrices

$$\begin{pmatrix} a_{11} & \cdots & a_{1i} \\ \cdot & \cdots & \cdot \\ \cdot & \cdots & \cdot \\ \cdot & \cdots & \cdot \\ a_{j1} & \cdots & a_{jj} \end{pmatrix}$$

are nonsingular, the matrix $A = (a_{ij})$ may be represented uniquely as the product $A = LDU$, where L is unit lower triangular, D is diagonal, and U is unit upper triangular. Any nonsingular matrix can be made to satisfy the conditions by a suitable rearrangement of rows and columns.

2. *LDU* **Theorem for Symmetric Matrices (Banachiewicz).** If A is symmetric and none of its upper-left principal submatrices is singular, then $A = LDL'$ where L is a unique unit lower triangular matrix, L' is its transpose, and D is diagonal. In fact, we may write $A = (LD^{\frac{1}{2}})(D^{\frac{1}{2}}L') = MM'$. However, unless A is also positive definite, M need not be real when A is real.

3. *LDU* **Theorem for Hermitian Matrices.** If H is Hermitian and none of its upper-left principal submatrices is singular, then $H = L(L^*)'$ where L is a unique lower triangular matrix and $(L^*)'$ is its conjugate transpose.

4. Every nonsingular matrix A is uniquely expressible as a product of a unitary matrix V by a positive-definite Hermitian matrix H, that is, $A = VH$. Every nonsingular *real* matrix, A, can be expressed as a product $A = SR$ where S is symmetric positive definite and R is orthogonal.

5. Let A be any real matrix. Then A^{-1} can be expressed as A' times the inverse, S^{-1}, of a symmetric positive-definite matrix S. Take $S = A'A$ which is both symmetric and positive definite. Then $A^{-1} = S^{-1}A'$.

6. If A is nonsingular, there exist two unitary matrices V_1 and V_2 such that $V_1AV_2 = \text{diag}(\mu_1, \ldots, \mu_n)$, where $\mu_1, \ldots, \mu_n$ are the positive square roots of the eigenvalues of $A(A^*)'$.

7. If A is tridiagonal and positive definite, then A can be written in the form $A = MM'$, where M is a bidiagonal matrix whose only nonzero elements are on the main diagonal and the adjacent diagonal below.

MATRIX TRANSFORMATION TO DIAGONAL FORM

8. Any matrix, A, is equivalent to a diagonal matrix. If A is transformed by a series of elementary row operations to the identity matrix I, then the result of applying the same operations to I (or to B) is the matrix A^{-1} (or $A^{-1}B$).

9. Any real symmetric matrix A is similar (and congruent) to a real diagonal matrix D by means of an orthogonal transformation R: $D = R^{-1}AR = R'AR$. Then the ith diagonal element of D may be taken as the ith eigenvalue of A and the ith column of R is then the ith eigenvector of A.

10. Any Hermitian matrix H is similar (and conjunctive) to a real diagonal matrix D by means of a unitary transformation V: $D = V^{-1}HV = (V^*)'HV$. The eigenvalues of H appear along the diagonal of D and the eigenvectors as corresponding columns of V.

11. Toeplitz Theorem. A necessary and sufficient condition for a matrix A to be similar to a diagonal matrix by means of a unitary transformation is that A be normal.

12. Any $n \times n$ matrix A which has n distinct eigenvalues is similar to a diagonal matrix. Neither the matrix effecting the transformation, nor the diagonal matrix need be real when A is real. A nondefective matrix in the complex field is similar to a diagonal matrix.

MATRIX TRANSFORMATION TO TRIANGULAR FORM

13. Any matrix is equivalent to a triangular matrix.

14. Schur's Theorem. Any matrix A is similar to a triangular matrix T by means of a unitary transformation V: $T = V^{-1}AV = (V^*)'AV$. The eigenvalues of A appear along the main diagonal of T. The elements of V may be complex if either A or its eigenvalues are complex.

MATRIX TRANSFORMATION TO TRIDIAGONAL FORM

15. An arbitrary matrix may be reduced to a similar tridiagonal matrix.

16. A real symmetric matrix, A, may be reduced by similarity (and congruence) to a tridiagonal symmetric matrix, $B = R'AR = R^{-1}AR$, where R is orthogonal. If the first column of R is the first column of the identity matrix, then both R and B are unique except for the signs of the off-diagonal elements.

MATRIX TRANSFORMATION TO COMPANION MATRIX

17. Every nonderogatory matrix is similar to the companion matrix of its characteristic equation.

MATRIX TRANSFORMATION TO JORDAN CANONICAL FORM

18. Any complex matrix is similar to a matrix in Jordan canonical form. Any real matrix is similar to a matrix in Jordan canonical form, but this form as well as the transforming matrix may have complex elements.

MATRIX TRANSFORMATION TO UNITARY OR ORTHOGONAL MATRIX

19. Schmidt's Theorem. In the complex field there exists a nonsingular triangular matrix which transforms any given nonsingular matrix into a unitary matrix. Within the real field such a transformation turns a nonsingular matrix into an orthogonal matrix.

POSITIVE-DEFINITE MATRICES

20. An $n \times n$ matrix $A = (a_{ij})$ is positive definite if and only if all the upper left-hand determinants

$$\begin{vmatrix} a_{11} & \cdots & a_{1j} \\ \cdot & & \cdot \\ \cdot & & \cdot \\ \cdot & & \cdot \\ a_{j1} & & a_{jj} \end{vmatrix} > 0$$

for $j = 1, \ldots, n$, and negative definite if and only if these determinants alternate in sign starting with $-$.

21. A matrix is positive definite if and only if its eigenvalues are all positive. It is negative definite if and only if its eigenvalues are all negative.

22. A real symmetric matrix A is positive definite if and only if there exists a real nonsingular triangular matrix P such that $A = PP'$.

EIGENVALUES AND CHARACTERISTIC EQUATIONS

23. Cayley-Hamilton Theorem. Any matrix satisfies its characteristic equation. If $P(\lambda) = |A - \lambda I|$, then $P(A) = 0$.

24. Similar matrices have the same characteristic and minimal polynomials and the same eigenvalues.

25. The eigenvalues of AB are the same as those of BA.

26. All eigenvalues of a real symmetric matrix or of a Hermitian matrix are real.

27. The eigenvalues of a unitary matrix have absolute value 1.

28. The eigenvalues of a real *skew symmetric* matrix are either zero or pure imaginaries, conjugate in pairs. If the matrix is of odd order, at least one eigenvalue is zero. The eigenvalues of a *skew Hermitian* matrix are all pure imaginaries.

29. The eigenvalues of an orthogonal matrix occur in reciprocal pairs.

30. Let A be an $n \times n$ complex matrix. Then $A^r \to 0$ as $r \to \infty$ if and only if each eigenvalue of A is of absolute value less than 1.

31. The identity $(I - C)^{-1} = I + C + C^2 + \cdots$ is valid if and only if each eigenvalue of C is less than 1 in absolute value.

32. Geršgorin Circle Theorem. The eigenvalues of A lie in the union of the circles

$$|z - a_{ii}| \leq \sum_{\substack{j=1 \\ j \neq i}}^{n} |a_{ij}| , \qquad i = 1, \ldots, n.$$

33. If the sums

$$\sum_{j=1}^{n} |a_{ij}| , \qquad i = 1, \ldots, n$$

are all less than 1, or if

$$\sum_{i=1}^{n} |a_{ij}| , \qquad j = 1, \ldots, n$$

are all less than 1, then all of the eigenvalues of A are inside the unit circle.

34. Bounds for eigenvalues of a matrix $A = (a_{ij})$. Let $\lambda_{\max}$ be the eigenvalue of A of greatest absolute value. Then the following inequalities hold:

1. $|\lambda_{\max}| \leq \sum_{i,k} |a_{ik}|$
2. $|\lambda_{\max}| \leq \left(\sum_{i,k} |a_{ik}|^2\right)^{1/2}$
3. $|\lambda_{\max}| \leq \max_{i,k} [n\ |a_{ik}|]$
4. $|\lambda_{\max}| \leq \max_{i} \sum_{k=1}^{n} |a_{ik}|$
5. $|\lambda_{\max}| \leq \max_{k} \sum_{i=1}^{n} |a_{ik}|$
6. $|\operatorname{Re}(\lambda_{\max})| \leq \max_{i,k} (\tfrac{1}{2}n)|\ a_{ik} + \bar{a}_{ki}|$
7. $|\operatorname{Im}(\lambda_{\max})| \leq \max_{i,k} (\tfrac{1}{2}n)\ |a_{ik} - \bar{a}_{ik}|$

35. The eigenvalues of A' are the same as those of A. Any eigenvector of A is orthogonal to any eigenvector of A' except its own pair.

36. The eigenvalues of A^{-1} are reciprocals of those of A. The eigenvectors of A and A^{-1} are the same.

37. If λ is an eigenvalue of A, then $\bar{\lambda}$ is an eigenvalue of A^* and of $(A^*)'$. The eigenvalues of $A(A^*)'$ are *not* in general $\lambda\bar{\lambda}$ unless A is a normal matrix.

$$\lambda_{\min}(A(A^*)') \leq |\lambda_i(A)|^2 \leq \lambda_{\max}(A(A^*)')$$

38. If λ is an eigenvalue of A, then λ^n is an eigenvalue of A^n.

39. If A is a real symmetric matrix, its eigenvectors form an orthogonal set. Even when multiple eigenvalues occur, the associated eigenvectors may be made orthogonal (and thus linearly independent).

40. If A is not symmetric but has distinct eigenvalues then its eigenvectors are linearly independent, but not necessarily orthogonal. If A has multiple eigenvalues, then its eigenvectors are not necessarily linearly independent.

41. The sum of the eigenvalues of A is equal to the trace of A, that is, trace $A = \Sigma\, \lambda_i$, and trace $A^m = \Sigma\, \lambda_i^{\,m}$.

42. A linear combination of eigenvectors associated with the same eigenvalue is also an eigenvector associated with that same eigenvalue. In particular if x is an eigenvector, then cx, where $c \neq 0$ is a constant, is an eigenvector. Linearly independent eigenvectors associated with the same eigenvalue can be chosen orthogonal.

43. If A has the eigenvalue λ and eigenvector x, then $P(A)$, where P is a rational function, has the eigenvalue $P(\lambda)$ and the eigenvector x. Also trace $P(A) = \Sigma\, P(\lambda_i)$.

44. If A has the eigenvalue λ and eigenvector x, then TAT^{-1} has the eigenvector Tx and eigenvalue λ.

DETERMINANTS d(A)

45. If each element of a row (column) is multiplied by its cofactor, the sum of these n products is the value of the determinant.

46. $\mathrm{d}(A') = \mathrm{d}(A)$.

47. $\mathrm{d}(B) = -\mathrm{d}(A)$ when B is obtained from A by the interchange of two rows or of two columns.

48. $\mathrm{d}(AB) = \mathrm{d}(A)\mathrm{d}(B)$.

49. $\mathrm{d}(B) = k[\mathrm{d}(A)]$ when B is obtained from A by multiplying any row or any column of A by k.

50. $\mathrm{d}(A) = 0$ if two rows or two columns are identical.

51. $\mathrm{d}(B) = \mathrm{d}(A)$ when B is obtained from A by adding to any row (or column) a linear combination of the other rows (columns).

MISCELLANEOUS

52. Rank is preserved by an equivalence transformation.

53. If in A a multiple of column i is added to column k, then in A^{-1} the same or another multiple of row k is subtracted from row i according as $i \neq k$ or $i = k$.

54. The product of any number of unitary (orthogonal) matrices is a unitary (orthogonal) matrix. The product of two Hermitian (symmetric) matrices is not in general Hermitian (symmetric).

55. The inverse of a symmetric (orthogonal, unitary, Hermitian) matrix is symmetric (orthogonal, unitary, Hermitian).

56. The inverse of an upper triangular (lower triangular, diagonal) matrix is upper triangular (lower triangular, diagonal).

57. The transpose of an upper triangular (lower triangular, diagonal) matrix is lower triangular (upper triangular, diagonal).

58. $(A + PBQ')^{-1} = A^{-1} - A^{-1}PB(B + BQ'A^{-1}PB)^{-1}BQ'A^{-1}$;
$(A + PB^{-1}Q')^{-1} = A^{-1} - A^{-1}P(B + Q'A^{-1}P)^{-1}Q'A^{-1}$.

59. $(AB)^{-1} = B^{-1}A^{-1}$.

60. $(A + B)' = A' + B'$; $(AB)' = B'A'$.

61. $(A + B)^* = A^* + B^*$; $(AB)^* = A^*B^*$.

62. $(A')^{-1} = (A^{-1})'$; $(A^*)^{-1} = (A^{-1})^*$; $(A^*)' = (A')^*$.

63. Schwarz' inequality: $|(x, y)| \leq \|x\| \cdot \|y\|$.

64. Properties of scalar product:

1. $(x, y) = (y, x)$ for x and y real vectors
2. $(x, y) = (\bar{y}, \bar{x})$ for x and y complex vectors
3. $(\alpha x + \beta y, z) = \alpha(x, z) + \beta(y, z)$ for vectors x, y, and z and scalars α and β.
4. $(x, x) \geq 0$ is real and equal to zero if and only if the vector x is zero.
5. $(x, Ay) = (Ax, y)$ if and only if A is a real symmetric or Hermitian matrix.
6. $(x, Ay) = (A'x, y)$ for x, y, A real.
7. $(x, Ay) = ((A^*)'x, y)$ for x, y, A complex.

65. Vector and matrix norms.

1. Vector norms $\|x\|$: Any vector norm must satisfy the following conditions:

(a) $\|x\| > 0$ unless $x = 0$
(b) $\|kx\| = |k|\,\|x\|$ where k is a complex constant
(c) $\|x + y\| \leq \|x\| + \|y\|$

Common vector norms are

$$\|x\|_E = \left(\sum_{i=1}^{n} |x_i|^2\right)^{1/2} \quad \text{the Euclidean norm or length}$$

$$\|x\|_\infty = \max_i |x_i|$$

$$\|x\|_1 = \sum_{i=1}^{n} |x_i| \,.$$

2. Matrix norms $\|A\|$. With any vector norm can be associated a matrix norm defined by

$$\|A\| = \max_{x \neq 0} \frac{\|Ax\|}{\|x\|}, \quad \text{or equivalently} \quad \|A\| = \max_{\|x\|=1} \|Ax\| \,.$$

Such a norm is called the matrix norm *subordinate* to the vector norm, and satisfies the following conditions:

(a) $\|A\| > 0$ unless $A = 0$
(b) $\|kA\| = |k| \, \|A\|$ where k is a complex constant
(c) $\|A + B\| \leq \|A\| + \|B\|$
(d) $\|AB\| \leq \|A\| \cdot \|B\|$

Any matrix and vector norms for which $\|Ax\| \leq \|A\| \, \|x\|$ holds are *consistent* or *compatible*. Clearly a vector norm and the subordinate matrix norm are compatible. Note that $\|I\| = 1$ for any subordinate matrix norm. Examples of matrix norms are

$$\|A\|_1 = \max_j \sum_{i=1}^{n} |a_{ij}| \quad \text{subordinate to} \quad \|x\|_1$$

$$\|A\|_\infty = \max_i \sum_{j=1}^{n} |a_{ij}| \quad \text{subordinate to} \quad \|x\|_\infty$$

Spectral norm $\|A\|_S = (\text{maximum eigenvalue of } A^{*\prime}A)^{1/2}$ subordinate to $\|x\|_E$.
Euclidean norm

$$\|A\|_E = \left(\sum_{i,j} |a_{ij}|^2\right)^{1/2} = [\text{Trace}\,(A^{*\prime}A)]^{1/2}$$

is consistent with $\|x\|_E$ but not subordinate to any vector norm since $\|I\|_E = \sqrt{n}$.
Absolute norm $\|A\|_A = \max_{i,j} |a_{ij}|$, also called the "maximum coefficient" of the matrix A.

The first three matrix-norm conditions (a), (b), (c) are satisfied by $\|A\|_1$, $\|A\|_\infty$, $\|A\|_S$, $\|A\|_E$, and $\|A\|_A$. Condition (d) is satisfied by $\|A\|_1$, $\|A\|_\infty$, $\|A\|_S$, and $\|A\|_E$. For $\|A\|_A$ we have instead of (d),

$$\|AB\|_A \leq n \, \|A\|_A \, \|B\|_A.$$

Also for the transpose A' of A, $\|A'\|_{A,E,S} = \|A\|_{A,E,S}$ whereas $\|A'\|_1 = \|A\|_\infty$ and $\|A'\|_\infty = \|A\|_1$.

The following relations hold between the norms:

$$\|AB\|_E \leq \|A\|_E \, \|B\|_S,$$

$$\|AB\|_E \leq \|A\|_S \, \|B\|_E,$$

$$\|A\|_A \leq \|A\|_S \leq \|A\|_E \leq n \, \|A\|_A,$$

$$\|A\|_E \leq \sqrt{n} \|A\|_S,$$

$$\| \, |A| \, \|_S \leq \| \, |A| \, \|_E = \|A\|_E \leq \sqrt{n} \, \|A\|_S,$$

where $|A|$ here indicates the matrix whose elements are the absolute values of the elements of A, that is, $|A| = (|a_{ij}|)$. Matrix norms provide an upper bound for the absolute values of the eigenvalues, for

$$Ax = \lambda x,$$

$$|\lambda| \, \|x\| = \|\lambda x\| = \|Ax\| \leq \|A\| \, \|x\|,$$

$$|\lambda| \leq \|A\|.$$

Thus

$$\|A\|_S{}^2 = (\text{maximum eigenvalue of } A^{*\prime}A) \leq \|A^{*\prime}A\| \leq \|A^{*\prime}\| \, \|A\|$$

$$\|A\|_S{}^2 \leq \|A^{*\prime}\|_\infty \, \|A\|_\infty = \|A\|_1 \, \|A\|_\infty.$$

66. Inverse of a Permuted Matrix. Let P_{ij} be the elementary matrix operator which interchanges rows i and j when used as a left multiplier and interchanges columns i and j when used as a right multiplier. Let

$$\tilde{A} = P_{i_nj_n} \cdots P_{i_1j_1}(A)P_{k_1l_1} \cdots P_{k_ml_m}.$$

Then

$$\tilde{A}^{-1} = P_{k_ml_m}^{-1} \cdots P_{k_1l_1}^{-1}(A^{-1})P_{i_1j_1}^{-1} \cdots P_{i_nj_n}^{-1}$$

and

$$A^{-1} = P_{k_1l_1} \cdots P_{k_ml_m}(\tilde{A}^{-1})P_{i_nj_n} \cdots P_{i_1j_1}.$$

67. If for every i,

$$2\,|a_{ii}| > \sum_j |a_{ij}|,$$

then the matrix A is nonsingular.

68. Newton Identities. Let

$$P(x) = x^n + \sum_{j=1}^n p_j x^{n-j} = 0$$

be the characteristic equation of A, and let

$$S_r = \text{trace } A^r = \sum_{i=1}^n \lambda_i^r.$$

Then

$$p_i = -\frac{1}{i}\left[\sum_{j=1}^i p_{i-j}S_j\right]$$

for $i = 1, \ldots, n$, with $p_0 = 1$.

69. Relation Between Inverse and Eigenvalue. Let C be the matrix whose columns are the eigenvectors of A. Let λ be the diagonal matrix of eigenvalues of A in the corresponding order. Then

$$\begin{aligned} AC &= C\lambda, \\ A &= C\lambda C^{-1}, \\ A^{-1} &= C\lambda^{-1}C^{-1} = C\lambda^{-1}C'. \end{aligned}$$

70. Let $v_1, v_2, \ldots, v_n$ be a set of n linearly independent n-component vectors. Then any arbitrary n-component vector x may be written uniquely as a linear combination of the vectors v_i, that is,

$$x = \sum_{i=1}^{n} \alpha_i v_i,$$

where the α_i are scalars.

APPENDIX C

TEST MATRICES

The measures of condition, $P(A)$, $M(A)$, and $N(A)$ are defined in Chapter 4. The determinant of A is written $\mathrm{d}(A)$ or $\det(A)$.

INVERSES GIVEN

1. Complex Matrix.

$$A_1 = \begin{pmatrix} 1 & 1+2i & 2+10i \\ 1+i & 3i & -5+14i \\ 1+i & 5i & -8+20i \end{pmatrix}$$

$$A_1^{-1} = \begin{pmatrix} 10+i & -2+6i & -3-2i \\ 9-3i & 8i & -3-2i \\ -2+2i & -1-2i & 1 \end{pmatrix}.$$

2. Matrix Due to T. S. Wilson [8, 27, 38]. A_2 is symmetric, ill-conditioned, and positive definite. $\mathrm{d}(A_2) = 1$, $M(A_2) = 2720$, $N(A_2) = 752$, $P(A_2) = 2984$.

$$A_2 = \begin{pmatrix} 5 & 7 & 6 & 5 \\ 7 & 10 & 8 & 7 \\ 6 & 8 & 10 & 9 \\ 5 & 7 & 9 & 10 \end{pmatrix}, \qquad b = \begin{pmatrix} 23 \\ 32 \\ 33 \\ 31 \end{pmatrix},$$

$$A_2^{-1} = \begin{pmatrix} 68 & -41 & -17 & 10 \\ -41 & 25 & 10 & -6 \\ -17 & 10 & 5 & -3 \\ 10 & -6 & -3 & 2 \end{pmatrix}, \qquad x = A_2^{-1}b = \begin{pmatrix} 1 \\ 1 \\ 1 \\ 1 \end{pmatrix}.$$

3. Special Case of Pei Matrix, A_{18}.

$$A_3 = (a_{ij}) \qquad A_3^{-1} = (\alpha_{ij})$$

$$a_{ij} = \begin{cases} 0 & \text{for} \quad i = j \\ 1 & \text{for} \quad i \neq j \end{cases} \qquad \alpha_{ij} = \begin{cases} -\dfrac{n-2}{n-1} & \text{for} \quad i = j \\ \dfrac{1}{n-1} & \text{for} \quad i \neq j, \end{cases}$$

where n is the order of A_3. A_3 is symmetric, but not positive definite.

4. Special Case of Pei Matrix, A_{18}.

$$A_4 = (a_{ij}) \qquad A_4^{-1} = (\alpha_{ij})$$

$$a_{ij} = \begin{cases} n & \text{for} \quad i = j \\ 1 & \text{for} \quad i \neq j \end{cases} \qquad \alpha_{ij} = \begin{cases} \dfrac{2}{2n-1} & \text{for} \quad i = j \\ -\dfrac{1}{(n-1)\,(2n-1)} & \text{for} \quad i \neq j, \end{cases}$$

where n is the order of A_4. The matrix is symmetric and positive definite for $n > 1$.

5. From [81].

$$A_5 = (a_{ij}) = \begin{cases} 2 & \text{for} \quad i = j \\ 1 & \text{for} \quad i \neq j \end{cases} \qquad A_5^{-1} = (\alpha_{ij}) = \begin{cases} \dfrac{n}{n+1} & \text{for} \quad i = j \\ \dfrac{-1}{n+1} & \text{for} \quad i \neq j. \end{cases}$$

A_5 is well-conditioned, symmetric, and positive definite, and is a special case of the Pei matrix, A_{18}.

6. From [81].

$$A_6 = (a_{ij}) = n - |i - j|$$

A_6 is well conditioned, symmetric, and positive definite.

$$A_6^{-1} = (\alpha_{ij}) = \begin{cases} \dfrac{n+2}{2n+2} & \text{if} \quad i = j = 1 \quad \text{or} \quad n \\ 1 & \text{if} \quad i = j \neq 1 \quad \text{or} \quad n \\ -\frac{1}{2} & \text{if} \quad |i - j| = 1 \quad \text{and} \quad n \neq 2 \\ -\frac{1}{3} & \text{if} \quad |i - j| = 1 \quad \text{and} \quad n = 2 \\ \dfrac{1}{2n+2} & \text{if} \quad |i - j| = n - 1 \neq 1 \\ 0 & \text{if} \quad 1 < |i - j| < n - 1 \end{cases}$$

7. Orthogonal Matrix [85].

$$A_7 = (a_{ij}) = \left(\frac{2}{n+1}\right)^{1/2} \sin\left(\frac{ij\pi}{n+1}\right),$$

where n is the order of A_7. $A_7^{-1} = A_7' = A_7$. A_7 is symmetric, orthogonal, *not* positive definite, has eigenvalues of absolute value 1, and $P(A_7) = 1$.

8. Givens' Matrix [85].

$$A_8 = (a_{ij}) = 2 \min(i, j) - 1 \quad \text{for} \quad i, j = 1, \ldots, n.$$

$$A_8^{-1} = \frac{1}{2}\begin{pmatrix} 3 & -1 & 0 & & & \\ -1 & 2 & -1 & & & \\ 0 & -1 & 2 & -1 & & \\ & & \cdots\cdots & & & \\ & & -1 & 2 & -1 & \\ & & 0 & -1 & 1 & \end{pmatrix} \quad \text{of order } n.$$

A_8 and A_8^{-1} are symmetric and positive definite; A_8^{-1} is tri-diagonal. $P(A_8) \sim 16n^2/\pi^2$.

9. From [8].

$$A_9 = \begin{pmatrix} 1 & -2 & 3 & 1 \\ -2 & 1 & -2 & -1 \\ 3 & -2 & 1 & 5 \\ 1 & -1 & 5 & 3 \end{pmatrix},$$

$$b = \begin{pmatrix} 3 \\ -4 \\ 7 \\ 8 \end{pmatrix}, \qquad x = A_9^{-1}b = \begin{pmatrix} 1 \\ 1 \\ 1 \\ 1 \end{pmatrix}.$$

A_k is symmetric, but not positive definite.

10. From [85].

$$A_{10} = (a_{ij}) = \begin{cases} \dfrac{i}{j} & \text{for} \quad i \le j \\ \dfrac{j}{i} & \text{for} \quad i > j \end{cases} \qquad i, j = 1, \ldots, n.$$

$$A_{10}^{-1} = (\alpha_{ij}) = \begin{cases} \dfrac{4i^3}{4i^2 - 1} & \text{for} \quad i = j, i < n \\ \dfrac{n^2}{2n - 1} & \text{for} \quad i = j = n \\ -\dfrac{i(i+1)}{2i+1} & \text{for} \quad j = i + 1 \\ -\dfrac{j(j+1)}{2j+1} & \text{for} \quad i = j + 1 \\ 0 & \text{for} \quad |i - j| > 1. \end{cases}$$

A_{10} and A_{10}^{-1} are symmetric and positive definite. A_{10}^{-1} is tridiagonal. All eigenvalues are real and distinct. $P(A_{10}) = cn^{H\epsilon}$, $0 \leq \epsilon \leq 1$.

11. Wilkinson [108].

$$A_{11} = \begin{pmatrix} 1 & 0 & 0 & 0 & 0 & 1 \\ 1 & 1 & 0 & 0 & 0 & -1 \\ -1 & 1 & 1 & 0 & 0 & 1 \\ 1 & -1 & 1 & 1 & 0 & -1 \\ -1 & 1 & -1 & 1 & 1 & 1 \\ 1 & -1 & 1 & -1 & 1 & -1 \end{pmatrix},$$

$$A_{11}^{-1} = \begin{pmatrix} 2^{-1} & 2^{-2} & -2^{-3} & 2^{-4} & -2^{-5} & 2^{-5} \\ 0 & 2^{-1} & 2^{-2} & -2^{-3} & 2^{-4} & -2^{-4} \\ 0 & 0 & 2^{-1} & 2^{-2} & -2^{-3} & 2^{-3} \\ 0 & 0 & 0 & 2^{-1} & 2^{-2} & -2^{-2} \\ 0 & 0 & 0 & 0 & 2^{-1} & 2^{-1} \\ 2^{-1} & -2^{-2} & 2^{-3} & -2^{-4} & 2^{-5} & -2^{-5} \end{pmatrix}.$$

A_{11} is a well-conditioned matrix which is not symmetric.

12. Hilbert Matrix [27, 38, 85, 92].

$$A_{12} = (a_{ij}) = \left(\frac{1}{i + j - 1}\right) \text{ for } i, j, = 1, \ldots, n.$$

$$A_{12}^{-1} = (\alpha_{ij}) = \frac{(-1)^{i+j}(n + i - 1)!\,(n + j - 1)!}{(i + j - 1)[(i - 1)!\,(j - 1)!]^2(n - i)!\,(n - j)!}$$

A_{12} is exponentially ill-conditioned and symmetric and $\log P(A_{12}) \sim Kn$, where $K \approx 3.5$; $M(A_{12}) \sim ke^{3.525n}$. For $n = 4$, $d(A_{12}) = 16.5344(10^{-7})$. Tables of inverses of A_{12} for orders $n = 2, \ldots, 10$ are given in [92].

13. From [38]. Define $A_{13} = A_{14}A_{14}'$. Then $A_{13}^{-1} = A_{14}'A_{14}$. A_{13} is ill-conditioned, symmetric and positive definite, and $\log P(A_{13}) \sim 4n \log 2$.

INVERSES AND EIGENVALUES GIVEN

14. Rutishauser [38].

$$A_{14} = \begin{pmatrix} 1 & 0 & 0 & 0 & 0 \\ 1 & -1 & 0 & 0 & 0 \\ 1 & -2 & 1 & 0 & 0 \\ 1 & -3 & 3 & -1 & 0 \\ 1 & -4 & 6 & -4 & 1 \end{pmatrix}$$

Since $A_{14}^2 = I$, $A_{14}^{-1} = A_{14}$. The eigenvalues and eigenvectors of A_{14} are

right: $v_1 = (6x_3 - 4x_4, 3x_3 - 2x_4, x_3, x_4, x_5)$ $\quad \lambda_i = 1, i = 1, 2, 3.$

$v_4 = (0, 2x_4 - x_5, 2x_4 - x_5, x_4, x_5)$ $\quad \lambda_i = -1, i = 4, 5.$

left: $u_1 = (y_1, y_2, y_5 - y_2, -2y_5, y_5)$ $\quad P(A_{14}) = 1.$

$u_4 = (y_1, \frac{1}{2}(y_4 - 4y_1), \frac{1}{2}(-3y_4), y_4, 0)$

15. From [81, 85].

$$A_{15} = (a_{ij}) = \begin{cases} -2 & \text{for } i = j \\ 1 & \text{for } |i - j| = 1 \\ 0 & \text{for } |i - j| > 1 \end{cases} \qquad i, j = 1, \ldots, n.$$

$$A_{15}^{-1} = (\alpha_{ij}) = \begin{cases} -i\dfrac{n - j + 1}{n + 1} & \text{for } i \le j \\ \alpha_{ji} & \text{for } i > j \end{cases}.$$

A_{15} is symmetric, tri-diagonal, and negative definite, so that $-A_{15}$ is positive definite. The eigenvalues of A_{15} are

$$\lambda_k = -4\sin^2\{k\pi/[2(n + 1)]\} \quad \text{for} \quad k = 1, \ldots, n.$$

Also $P(A_{15}) \sim 4n^2/\pi^2$ and, for $n = 4$, $d(A_{15}) = 5$. A_{15}^2 is positive definite and $P(A_{15}^2) \sim 16n^4/\pi^4$. A_{15}^3 is negative definite and $P(A_{15}^3) \sim 64n^6/\pi^6$.

16. From [85].

$$A_{16} = (a_{ij}) = \left(\frac{i + j}{p}\right) = \begin{cases} 0 & \text{if } p \text{ divides } i + j, \\ 1 & \text{if } i + j \text{ is congruent to a square mod } p, \\ -1 & \textit{otherwise.} \end{cases}$$

The order of A_{16} is $n = p - 1$ where p is an odd prime.

$$A_{16}^{-1} = (\alpha_{ij}) = \frac{1}{p}\left[\left(\frac{i + j}{p}\right) - \left(\frac{i}{p}\right) - \left(\frac{j}{p}\right)\right].$$

If

$$B = \begin{pmatrix} 0 & 1 & 0 & 0 & 0 \\ 0 & 0 & 1 & 0 & 0 \\ 0 & 0 & 0 & 1 & 0 \\ 0 & 0 & 0 & 0 & 1 \\ 10^{-5} & 0 & 0 & 0 & 0 \end{pmatrix}, \quad \text{then} \quad A_{23} = \begin{pmatrix} B & 2B \\ 4B & 3B \end{pmatrix}.$$

Eigenvalues of A_{23} are

$$0.5\left[\cos\left(\frac{2\pi k}{5}\right) + i \sin\left(\frac{2\pi k}{5}\right)\right], \qquad -0.1\left[\cos\left(\frac{2\pi k}{5}\right) + i \sin\left(\frac{2\pi k}{5}\right)\right]$$

for $k = 1, \ldots, 5$.

24. Circulant [27, 85].

$$A_{24} = \begin{pmatrix} a_1 & a_2 & \cdots & a_n \\ a_n & a_1 & \cdots & a_{n-1} \\ \cdot & & & \cdot \\ \cdot & & & \cdot \\ \cdot & & & \cdot \\ a_2 & a_3 & \cdots & a_1 \end{pmatrix}.$$

The eigenvalues of the circulant A_{24} are

$$\lambda_k = a_1 + a_2\zeta_k + a_3\zeta_k^2 + \cdots + a_n\zeta_k^{n-1}$$

where ζ_k runs over the nth roots of unity, that is,

$$\zeta_k = \cos\left(\frac{2\pi k}{n}\right) + i \sin\left(\frac{2\pi k}{n}\right), \quad \text{for} \quad k = 1, 2, \ldots, n.$$

The eigenvectors are

$$v_k = n^{-1/2}(\zeta_k^{n-1}, \zeta_k^{n-2}, \ldots, \zeta_k, 1) \quad \text{for} \quad k = 1, \ldots, n.$$

Special Circulants.

$B_2 =$ circulant $(t^2, 1, 2, \ldots, t, t + 1, t, t - 1, \ldots, 1)$, where $n = 2t + 2$.

$B_3 =$ circulant $(n + 1, 1, 1, \ldots, 1)$.

$B_5 =$ circulant $(1, 2, 3, \ldots, n)$.

$P(B_2) \sim 1.4$, $P(B_3) = 2$, $P(B_5) \sim n$.

25. Rutishauser [35].

$$A_{25} = \begin{pmatrix} 6 & 4 & 4 & 1 \\ 4 & 6 & 1 & 4 \\ 4 & 1 & 6 & 4 \\ 1 & 4 & 4 & 6 \end{pmatrix}.$$

Eigenvalues:

$$\lambda_1 = 15,$$
$$\lambda_2 = \lambda_3 = 5,$$
$$\lambda_4 = -1.$$

Eigenvectors:

$$v_1 = (1, 1, 1, 1),$$
$$v_3 = v_2 = (-x_4, -x_3, x_3, x_4),$$
$$v_4 = (1, -1, -1, 1).$$

26. Rutishauser [35].

$$A_{26} = \begin{pmatrix} 5 & 4 & 1 & 1 \\ 4 & 5 & 1 & 1 \\ 1 & 1 & 4 & 2 \\ 1 & 1 & 2 & 4 \end{pmatrix}.$$

Eigenvalues:

$$\lambda_1 = 10,$$
$$\lambda_2 = 5,$$
$$\lambda_3 = 2,$$
$$\lambda_4 = 1.$$

Eigenvectors:

$$v_1 = (2, 2, 1, 1)$$
$$v_2 = (-1, -1, 2, 2)$$
$$v_3 = (0, 0, -1, 1)$$
$$v_4 = (-1, 1, 0, 0).$$

27. Bodewig [35, 115, 123].

$$A_{27} = \begin{pmatrix} 2 & 1 & 3 & 4 \\ 1 & -3 & 1 & 5 \\ 3 & 1 & 6 & -2 \\ 4 & 5 & -2 & -1 \end{pmatrix}.$$

Eigenvalues:

$$\lambda_1 = 7.93290472,$$
$$\lambda_2 = -8.02857835,$$
$$\lambda_3 = 5.66886436,$$
$$\lambda_4 = 1.57319074.$$

Eigenvectors:

$$v_1 = (1, 0.37781815, 1.38662122, 0.34880573),$$
$$v_2 = (1, 2.50146030, -0.75773064, -2.56421169),$$
$$v_3 = (1, 0.95700152, -1.42046826, 1.74331693),$$
$$v_4 = (1, -0.90709211, -0.37759122, -0.38331238).$$

28. Rutishauser [35].

$$A_{28} = \begin{pmatrix} 4 & -5 & 0 & 3 \\ 0 & 4 & -3 & -5 \\ 5 & -3 & 4 & 0 \\ 3 & 0 & 5 & 4 \end{pmatrix}.$$

Eigenvalues:

$$\lambda_1 = 12,$$
$$\lambda_2 = 1 + 5i,$$
$$\lambda_3 = 1 - 5i,$$
$$\lambda_4 = 2.$$

Eigenvectors:

right: $v_1 = (1, -1, 1, 1),$
$v_4 = (1, 1, -1, 1),$
$v_2 = (1, -i, -i, -1),$
$v_3 = (1, i, i, -1).$

left: $u_1 = (1, -1, 1, 1),$
$u_4 = (1, 1, -1, 1),$
$u_2 = (1, i, i, -1),$
$u_3 = (1, -i, -i, -1).$

29. Striped matrix [35, 124]. $A_{29} = (a_{ij})$ is of order 11 and is symmetric.

$$a_{ij} = \begin{cases} 6 & \text{for} \quad i = j \\ 3 & \text{for} \quad |i-j| = 1 \\ 1 & \text{for} \quad |i-j| = 2 \\ 1 & \text{for} \quad |i-j| = 3 \\ 0 & \text{for} \quad |i-j| > 3 \end{cases}$$

except

$$a_{11} = a_{11,11} = 5,$$
$$a_{21} = a_{12} = a_{10,11} = a_{11,10} = 2.$$

Eigenvalues:

$$\lambda_1 = 14.941819327676382,$$
$$\lambda_2 = 12.196152422706632,$$
$$\lambda_3 = 8.8284271247461900,$$
$$\lambda_4 = 6,$$
$$\lambda_5 = 4.4066499006731521,$$
$$\lambda_6 = 4.1292484841890931,$$
$$\lambda_7 = 4,$$
$$\lambda_8 = 4,$$
$$\lambda_k = 3.1715728752538100,$$
$$\lambda_{10} = 1.8038475772933680,$$
$$\lambda_{11} = 0.52228228746137256.$$

30. Rutishauser [35]. $A_{30} = (a_{ij}),\ i, j = 1 \ldots, 89.$

$$a_{ij} = \begin{cases} 20 & \text{for} \quad i = j, \\ 15 & \text{for} \quad |i - j| = 1, \\ 6 & \text{for} \quad |i - j| = 2, \\ 1 & \text{for} \quad |i - j| = 3, \\ 0 & \text{for} \quad |i - j| > 3, \end{cases}$$

except

$$a_{11} = a_{12} = a_{21} = a_{89,89} = a_{88,89} = a_{89,88} = 14.$$

$A_{30} = T^3$ where

$$T = (t_{ij}) = \begin{cases} 2 & \text{for} \quad i = j \\ 1 & \text{for} \quad |i - j| = 1 \\ 0 & \text{for} \quad |i - j| > 1 \end{cases}$$

Eigenvalues of A_{30} are

$$\lambda_k = 64 \cos^6\left(\frac{k\pi}{2(n + 1)}\right) \quad \text{for} \quad k = 1, \ldots, 89.$$

A_{30} is a symmetric band matrix.

31. From [119]. Let

$$D = \begin{pmatrix} -2 & 2 & 2 & 2 \\ -3 & 3 & 2 & 2 \\ -2 & 0 & 4 & 2 \\ -1 & 0 & 0 & 5 \end{pmatrix}, \quad C = \begin{pmatrix} 6D & -D & D & 0 \\ 8D & 0 & D & 2D \\ -2D & 0 & D & 2D \\ 5D & -D & -D & D \end{pmatrix}, \quad B = \begin{pmatrix} 3C & 3C \\ 5C & C \end{pmatrix},$$

and

$$A_{31} = \begin{pmatrix} B & 2B \\ 4B & 3B \end{pmatrix}.$$

A_{31} is of order 64. Let $m = 3 \pm i,\ 1 \pm 2i$. Then the eigenvalues of A_{31} are

$$120m,\ -40m,\ -24m,\ 8m,\ 90m,\ -30m,\ -18m,\ 6m,\ 60m,$$
$$-20m,\ -12m,\ 4m,\ 30m,\ -10m,\ -6m,\ 2m.$$

32. Parlett [119].

$A_{32} = (a_{ij})$ for $i, j = 1, 100$.
$a_{ij} = 0$ for $i < j$ except $a_{12} = 40/102$, $a_{1,100} = 40$,
$a_{ii} = 101 - i$,
$a_{ij} = (-1)^{i+j+1}\, 40/(i + j - 2)$ for $i > j$.

Eigenvalues are 100, 99, . . . , 2, 1.

33. Lotkin [116].

$$A_{33} = \begin{bmatrix} 1 & 1 & 1 & 1 & 1 & 1 \\ \frac{1}{2} & \frac{1}{3} & \frac{1}{4} & \frac{1}{5} & \frac{1}{6} & \frac{1}{7} \\ \frac{1}{3} & \frac{1}{4} & \frac{1}{5} & \frac{1}{6} & \frac{1}{7} & \frac{1}{8} \\ \frac{1}{4} & \frac{1}{5} & \frac{1}{6} & \frac{1}{7} & \frac{1}{8} & \frac{1}{9} \\ \frac{1}{5} & \frac{1}{6} & \frac{1}{7} & \frac{1}{8} & \frac{1}{9} & \frac{1}{10} \\ \frac{1}{6} & \frac{1}{7} & \frac{1}{8} & \frac{1}{9} & \frac{1}{10} & \frac{1}{11} \end{bmatrix}.$$

The eigenvalues given by Lotkin are

$$0.2132376 \times 10,\quad -0.2214068,\quad -0.3184330 \times 10^{-1},$$
$$-0.8983233 \times 10^{-3},\quad -0.1706278 \times 10^{-4},\quad -0.1394499 \times 10^{-6}.$$

34. Wilkinson [116, 125].

$$A_{34} = \begin{pmatrix} 2+3i & 3+i & & & & & & & & \\ 3+2i & -2-i & 1+2i & & & & & & & \\ 5-3i & 1+2i & 2+i & -1+4i & & & & 0 & & \\ 2+6i & -2+3i & 3-i & -4+2i & 5+5i & & & & & \\ 1+4i & 2+2i & -3+7i & 1+5i & 2-3i & 1+6i & & & & \\ 5-i & 0+4i & 1+5i & -8-i & 4+7i & 7+i & 4-2i & & & \\ 5+2i & 1+4i & 6-5i & 8+4i & 4-4i & -1+5i & 3+0i & -4+6i & & \\ -4-3i & 7+3i & 1+6i & 2-4i & 3+i & 1+2i & 1+4i & 6+3i & 7-i & \\ 5+0i & 2+2i & 1+3i & 1+i & -4-2i & 1+6i & 1+2i & 2+5i & 0+i & 3+2i \\ 5+2i & 2+6i & 1-3i & 7+4i & 4+i & -7+0i & 3-3i & 5-4i & 6+3i & 2+5i \end{pmatrix}.$$

Eigenvalues:

$$\begin{array}{r}
4.16174868 + 3.13751356i \\
5.43644837 - 3.97142582i \\
2.38988759 + 7.26807071i \\
-1.93520144 - 3.97509382i \\
-2.44755082 + 0.437126175i \\
-5.27950616 - 2.27596303i \\
1.03205812 + 9.29413278i \\
-4.96687009 - 8.08712475i \\
8.81130928 + 1.54938266i \\
10.7976764 \quad + 8.62338151i.
\end{array}$$

Hessenberg matrix of order 10 which is well conditioned with respect to eigenvalue computation.

35. Rosser matrix [120].

$$A_{35} = \begin{pmatrix}
611 & 196 & -192 & 407 & -8 & -52 & -49 & 29 \\
 & 899 & 113 & -192 & -71 & -43 & -8 & -44 \\
 & & 899 & 196 & 61 & 49 & 8 & 52 \\
 & & & 611 & 8 & 44 & 59 & -23 \\
 & & & & 411 & -599 & 208 & 208 \\
 & \text{symmetric} & & & & 411 & 208 & 208 \\
 & & & & & & 99 & -911 \\
 & & & & & & & 99
\end{pmatrix}.$$

Eigenvalues (let $a = \sqrt{10405}$, $b = \sqrt{26}$):

$$\begin{array}{ll}
1020.04901843 & = 10a \\
1020.0000 & \\
1019.90195136 & = 510 + 100b \\
1000. & \\
1000. & \\
0.09804864072 & = 510 - 100b \\
0.0 & \\
-1020.04901843 & = -10a.
\end{array}$$

Trace is 4040.0.

Eigenvectors (let $a = \sqrt{10405}$, $b = \sqrt{26}$):

$$\begin{aligned}
&(2, 1, 1, 2, 102 - a, 102 - a, -204 + 2a, -204 + 2a),\\
&(1, -2, -2, 1, 2, -2, 1, -1),\\
&(2, -1, 1, -2, 5 + b, -5 - b, -10 - 2b, 10 + 2b),\\
&(1, -2, -2, 1, -2, 2, -1, 1),\\
&(7, 14, -14, -7, -2, -2, -1, -1),\\
&(2, -1, 1, -2, 5 - b, -5 + b, -10 + 2b, 10 - 2b),\\
&(1, 2, -2, -1, 14, 14, 7, 7),\\
&(2, 1, 1, 2, 102 + a, 102 + a, -204 - 2a, -204 - 2a).
\end{aligned}$$

36. Ortega's Matrices [88]. Let $C = I + uv'$, where I is the $n \times n$ identity matrix and u and v are n-component column vectors. Instead of v' the conjugate transpose $(v^*)'$ may also be used. Then $C^{-1} = I - (1 + v'u)^{-1}uv'$. Define $A_{36} = CRC^{-1}$, where R is a matrix whose eigenvalues, eigenvectors and inverse are known. Then $A_{36}{}^{-1} = CR^{-1}C^{-1}$. Any vector orthogonal to v is a characteristic vector of C corresponding to the eigenvalue $\lambda = 1$, and u is a characteristic vector corresponding to $\lambda = 1 + v'u$. Since A_{36} and R are similar, the eigenvalues of A_{36} are the same as those of R, and the columns of C are the eigenvectors of A.

Examples.

1. Choose v such that

$$\sum_{i=1}^{n} v_i{}^2 = 1, \text{ say } v_i = \frac{1}{\sqrt{n}}.$$

Let $u = -2v$. Let $R = \text{diag}\,(d_1, \ldots, d_n)$. Then $C = I - 2vv' = C^{-1}$, $A_{36} = CRC$, $A_{36}{}^{-1} = CR^{-1}C$, where $R^{-1} = \text{diag}\,(1/d_1, \ldots, 1/d_n)$. A_{36} is symmetric with eigenvalues $d_1, \ldots, d_n$ and eigenvectors the columns of C.

For example, generate the real symmetric matrix A_{36} of order n: Define

$$d_i = i,$$

$$R = \text{diag}\,(d_1, d_2, \ldots d_n),$$

$$v_i = \frac{1}{\sqrt{n}}$$

$$C = C^{-1} = C' = I - 2vv',$$

$$C = (c_{ij}) \quad \text{where} \quad \begin{cases} c_{ii} = 1 - \dfrac{2}{n} & \text{for all } i,\\[2ex] c_{ij} = -\dfrac{2}{n} & \text{for all } i, j \text{ such that } i \neq j. \end{cases}$$

$$A_{36} = CRC.$$

Any element a_{ps} of A_{36} may be written as

$$a_{ps} = \sum_{q=1}^{n} d_q c_{pq} c_{qs} = \sum_{q=1}^{n} q c_{pq} c_{qs}.$$

If $p \neq s$, then

$$a_{ps} = \sum_{\substack{q=1\\ q\neq p,s}}^{n} q\left(\frac{4}{n^2}\right) - (p+s)\left(\frac{2}{n}\right)\left(1-\frac{2}{n}\right)$$

$$= \frac{4}{n^2}\left[\frac{n(n+1)}{2} - (p+s)\right] - \frac{2}{n^2}\left[(n-2)(p+s)\right],$$

$$a_{ps} = \frac{2}{n}[(n+1)-(p+s)].$$

If $p = s$, then

$$a_{pp} = \sum_{\substack{q=1\\ q\neq p}}^{n} q\left(\frac{4}{n^2}\right) + p\left(1-\frac{2}{n}\right)^2 = \frac{4}{n^2}\left[\frac{n(n+1)}{2} - p\right] + \frac{p}{n^2}(n-2)^2,$$

$$a_{pp} = \frac{1}{n}[2(n+1) + p(n-4)].$$

Hence A_{36} may be generated easily using the expressions for a_{ps} and a_{pp}. The eigenvalues of A_{36} are $\lambda_i = i$ and the eigenvectors are the columns of the matrix C. These eigenvectors are normalized. Positions $(k, n-k+1)$, for $k = 1, \ldots, n$, of the matrix A_{36} are zero, so that when n is odd A_{36} has a zero diagonal. The matrix C is symmetric and orthogonal.

2. Choose $R = \text{diag}\,(d_1, \ldots, d_n)$, where $n = 2k$ is even. Choose $u' = \alpha(1, 1, \ldots, 1)$, where α is a scalar.

$$v' = (\underbrace{1, \ldots, 1}_{k}, \underbrace{-1, \ldots, -1}_{k}).$$

Then $C = I + uv'$, $C^{-1} = I - (1 + v'u)^{-1}uv' = I - uv'$, $A_{36} = CRC^{-1}$. A_{36} is *not* symmetric. The parameter α allows some control over "condition". Note that $u'v = v'u = 0$. Define the condition number of the mth eigenvalue to be the Euclidean length of the mth row of C^{-1} times the length of the mth column of C, that is, if $C = (c_{ij})$ and $C^{-1} = (\gamma_{ij})$, then

$$\text{condition number} = \left(\sum_{j=1}^{n} \gamma_{mj}^{\,2}\right)^{1/2}\left(\sum_{i=1}^{n} c_{im}^{\,2}\right)^{1/2}$$

$$= [n^2\alpha^4 + 2(n-2)\alpha^2 + 1]^{1/2}.$$

As an example generate the nonsymmetric complex matrix A_{36C} of order 4. Define

$$\begin{aligned} d_r &= r + (4 + r)i \\ R &= \mathrm{diag}(d_1, d_2, d_3, d_4) \\ u' &= (1, 1, 1, 1) \\ v' &= (1, 1, -1, -1) \\ C &= I + uv' \\ C^{-1} &= I - uv' \\ A_{36C} &= CRC^{-1}. \end{aligned}$$

The resulting matrix and its eigenvalues and vectors are

$$A_{36C} = \begin{pmatrix} 5+9i & 5+5i & -6-6i & -7-7i \\ 3+3i & 6+10i & -5-5i & -6-6i \\ 2+2i & 3+3i & -1+3i & -5-5i \\ 1+i & 2+2i & -3-3i & 0+4i \end{pmatrix}.$$

$\lambda_1 = 1 + 5i$, $\lambda_2 = 2 + 6i$, $\lambda_3 = 3 + 7i$, $\lambda_4 = 4 + 8i$.

$$x^{(1)} = \begin{pmatrix} 2 \\ 1 \\ 1 \\ 1 \end{pmatrix}, \quad x^{(2)} = \begin{pmatrix} 1 \\ 2 \\ 1 \\ 1 \end{pmatrix}, \quad x^{(3)} = \begin{pmatrix} -1 \\ -1 \\ 0 \\ -1 \end{pmatrix}, \quad x^{(4)} = \begin{pmatrix} -1 \\ -1 \\ -1 \\ 0 \end{pmatrix}.$$

3. Choose $R = \mathrm{diag}\,(d_1, \ldots, d_n)$, where $n = 2k$ is even;

$$\begin{aligned} u' &= (1, 2, \ldots, k, 1, 2, \ldots, k), \\ v' &= (1, 2, \ldots, k, -1, -2, \ldots, -k). \end{aligned}$$

Again $u'v = v'u = 0$. The condition number of the mth and $(m + k)$th eigenvalues is

$$\tfrac{1}{3}\{[3 + m^2k(k + 1)(2k + 1)]^2 - 36m^4\}^{1/2}.$$

37. Frank [41, 119, 125, 126].

$$A_{37} = \begin{pmatrix} 1 & 1 & & & & & \\ 1 & 2 & 2 & & & & \\ 1 & 2 & 3 & 3 & & & \\ 1 & 2 & 3 & 4 & 4 & & \\ & \cdots & & & & & \\ 1 & 2 & 3 & 4 & \cdots & (n-1) & (n-1) \\ 1 & 2 & 3 & 4 & \cdots & (n-1) & n \end{pmatrix}.$$

Det(A_{37}) = 1 for any n. The value of the determinant is sensitive to changes in elements near the bottom left corner. The values of the condition measure $v_i'u_i = s_i$ for normalized eigenvectors u_i of A and v_i of A' are very small for small eigenvalues and are near 1 for larger ones. Eigenvalues for $n = 12$: $\lambda =$

0.03102 80606 44010
0.04950 74291 85278
0.08122 76592 40405
0.14364 65197 69220
0.28474 97205 58478
0.64350 53190 04855 5
1.55398 87091 32107 90
3.51185 59485 80757 194
6.96153 30855 67122 113
12.31107 74088 68526 120
20.19898 86458 77079 428
32.22889 15015 72160 750.

38. Wilkinson [125].

Eigenvalues are

2.06853152 − 2.05443045i
2.40341933 + 2.08105512i
2.72491267 − 2.37837845i
2.45640400 + 0.631936861i
2.27740066 + 1.44826850i
0.812811959 + 1.33551135i
−1.38565721 − 1.38756051i
−2.72480368 + 0.657064546i
1.57598142 − 3.83032770i
3.28048252 + 3.27566163i
1.19252750 − 5.44399752i
3.55339888 + 1.26465631i
−2.45560768 − 4.69290496i
−4.89673115 + 3.62210856i
−5.65716067 + 1.63200082i
5.77408994 + 2.83933591i

A_{38} is well conditioned with respect to eigenvalues.

$$A_{38} = \begin{pmatrix}
(3+2i) & (4-i) & & & & & & & & & & & & & & \\
(4-i) & (-1-i) & (2+4i) & & & & & & & & & & & & & \\
 & (2+4i) & (3-4i) & (3+i) & & & & & & & & & & & & \\
 & & (3+i) & (2+3i) & (3-2i) & & & & & & & & & & & \\
 & & & (3-2i) & (-5+i) & (2-2i) & & & & & & & & & & \\
 & & & & (2-2i) & (1+2i) & (2+3i) & & & & & & & & & \\
 & & & & & (2+3i) & (5+2i) & (1+3i) & & & & & & & & \\
 & & & & & & (1+3i) & (-2+i) & (-2+2i) & & & & & & & \\
 & & & & & & & (-2+2i) & (1-2i) & (3+3i) & & & & & & \\
 & & & & & & & & (3+3i) & (-1-4i) & (-1+5i) & & & & & \\
 & & & & & & & & & (-1+5i) & (2+i) & (4+3i) & & & & \\
 & & & & & & & & & & (4+3i) & (1-5i) & (1-6i) & & & \\
 & & & & & & & & & & & (1-6i) & (3+i) & (2+i) & & \\
 & & & & & & & & & & & & (2+i) & (2+4i) & (5-i) & \\
 & & & & & & & & & & & & & (5-i) & (-4+3i) & (-3-4i) \\
 & & & & & & & & & & & & & & (-3-4i) & (1-5i)
\end{pmatrix}$$

39. Frank [118].

$$A_{39} = \begin{pmatrix} 12 & 11 & 10 & 9 & \cdot & \cdot & \cdot & 2 & 1 \\ 11 & 11 & 10 & 9 & \cdot & \cdot & \cdot & 2 & 1 \\ 10 & 10 & 10 & 9 & \cdot & \cdot & \cdot & 2 & 1 \\ \cdot & \cdot & \cdot & \cdot & \cdot & \cdot & \cdot & \cdot & \cdot \\ \cdot & \cdot & \cdot & \cdot & \cdot & \cdot & \cdot & \cdot & \cdot \\ \cdot & \cdot & \cdot & \cdot & \cdot & \cdot & \cdot & \cdot & \cdot \\ 2 & 2 & 2 & 2 & \cdot & \cdot & \cdot & 2 & 1 \\ 1 & 1 & 1 & 1 & \cdot & \cdot & \cdot & 1 & 1 \end{pmatrix}.$$

Eigenvalues are

$$\lambda_i = \frac{1}{2}\left[1 - \cos\frac{(2i-1)\pi}{25}\right]^{-1} \quad \text{for} \quad i = 1, \ldots, 12.$$

Characteristic polynomial is

$$P(\lambda) = \lambda^{12} - 78\lambda^{11} + 1001\lambda^{10} - 5005\lambda^{9} + 12870\lambda^{8} - 19448\lambda^{7} + 18564\lambda^{6} - 11628\lambda^{5} + 4845\lambda^{4} - 1330\lambda^{3} + 231\lambda^{2} - 23\lambda + 1.$$

40. Wilkinson [41].

$$A_{40} = (a_{ij}),\ a_{ii} = i, \quad a_{i,i+1} = 20, \quad a_{ij} = 0 \text{ otherwise.}$$

Characteristic equation is

$$P(\lambda) = \prod_{i=1}^{20} (i - \lambda) = 0.$$

If $a_{20,1}$ is changed from zero to ϵ the characteristic equation becomes

$$\prod_{i=1}^{20} (i - \lambda) = 20^{19}\epsilon.$$

Each u_i is almost orthogonal to the corresponding v_i, where u_i is the ith eigenvector of A and v_i is the eigenvector of A'. The condition measure for $i = 10$ is $v'_{10}u_{10} = 9!\,10!/20^{19}$. The eigenvalues are well separated but ill-conditioned.

41. Eberlein [117]. $A_{41} = (a_{ij})$ where

$$\begin{aligned} a_{ii} &= -[(2i+1)N + is - 2i^2], \\ a_{i,i+1} &= (i+1)(N+s-i), \\ a_{i,i-1} &= i(N-i+1), \\ a_{ij} &= 0 \quad \text{for} \quad |i-j| > 1, \end{aligned}$$

and $i, j = 0, 1, \ldots, N$; s is an arbitrary parameter and $N + 1$ is the order of the matrix. The eigenvalues are

$$\lambda_j = -j(s + j + 1), \qquad j = 0, 1, \ldots, N.$$

Left eigenvectors are

$$v_i^{(j)} = \frac{1}{\binom{N}{i}} \sum_{k=0}^{q} (-1)^k \binom{N-k}{N-i} \binom{j}{k} \binom{s+j+k}{k} \quad \begin{array}{l} \text{for} \quad i = 0, \ldots, N \\ q = \min(i, j). \end{array}$$

Let n be an integer such that $1 \leq n \leq N$. The components of the right eigenvectors are

$$u_i^{(j)} = \binom{N+s-i}{N-i} v_i^{(j)} \quad \text{unless} \quad s = -n \quad \text{and} \quad j \geq n.$$

If $s = -n$ and $j \geq n$, then

$$u_i^{(j)} = \left[\frac{\binom{N+s-i}{N-i}}{(n+s)} \right] v_i^{(j)} \text{ for } \quad i \leq N - n,$$

$$u_i^{(j)} = \left[\frac{\binom{N+s-i}{N-i}}{(n+s)} \right] \frac{v_i^{(j)}}{(n+s)} \quad \text{for} \quad i > N - n.$$

When $s = -2, \ldots, -2N$, the matrix is defective with two or more pairs of eigenvectors coalescing.

REFERENCES

BOOKS

[1] Aitken, A. C., *Determinants and Matrices*, Interscience, New York, 1944.

[2] Alt, F. L., and M. Rubinoff (eds.), *Advances in Computers*, Vol. 3, Academic Press, New York, 1962.

[3] Beckenbach, E. F. (ed.), *Modern Mathematics for the Engineer*, McGraw-Hill, New York, 1956.

[4] Beckenbach, E. F. (ed.), *Modern Mathematics for the Engineer*, Second Series, McGraw-Hill, New York, 1961.

[5] Berezin, I. S., and N. P. Zhidkov, *Computing Methods*, Eng. Ed. Vol. II, translated by O. M. Blunn, Pergamon, London, 1965.

[6] Birkhoff, G., and S. MacLane, *A Survey of Modern Algebra*, Macmillan, New York, 1946.

[7] Bodewig, E., *Matrix Calculus*, North-Holland, Amsterdam, 1956.

[8] Booth, A. D., *Numerical Methods*, Butterworths, London, 1957.

[9] Curtiss, J. H. (ed.), *Am. Math. Soc., Numerical Analysis: Proceedings of Symposia in Applied Mathematics*, Vol. VI, McGraw-Hill, New York, 1956.

[10] Dwyer, P. S., *Linear Computations*, Wiley, New York, 1951.

[11] Faddeeva, V. N., *Computational Methods of Linear Algebra*, translated by Curtis D. Benster, Dover, New York, 1959.

[12] Forsythe, G. E., and W. R. Wasow, *Finite-Difference Methods for Partial Differential Equations*, Wiley, New York, 1960.

[13] Fox, L., *An Introduction to Numerical Linear Algebra*, Clarendon, Oxford, 1964.

[14] Frazer, R. A., W. J. Duncan, and A. R. Collar, *Elementary Matrices*, Cambridge University Press, London, 1947.

[15] Gantmacher, F. R., *Applications of the Theory of Matrices*, translated by J. L. Brenner, D. W. Bushaw, and S. Evanusa, Interscience, New York, 1959.

[16] Hammersley, J. M. and D. C. Handscomb, *Monte Carlo Methods*, Methuen, London (Wiley, New York), 1964.

[17] Henrici, P., *Elements of Numerical Analysis*, Wiley, New York, 1964.

[18] Hildebrand, F. B., *Introduction to Numerical Analysis*, McGraw-Hill, New York, 1956.

[19] Householder, A. S., *Principles of Numerical Analysis*, McGraw-Hill, New York, 1953.

[20] Householder, A. S., *The Theory of Matrices in Numerical Analysis*, Blaisdell, New York, 1964.
[21] Kunz, K. S., *Numerical Analysis*, McGraw-Hill, New York, 1957.
[22] Lance, G. N., *Numerical Methods for High-Speed Computers*, Iliffe, London, 1960.
[23] Lanczos, C., *Applied Analysis*, Prentice-Hall, Englewood Cliffs, New Jersey, 1961.
[24] Langer, R. E. (ed.), *Boundary Problems in Differential Equations*, University of Wisconsin Press, Madison, Wisconsin, 1960.
[25] MacDuffee, C. C., *Theory of Matrices*, Chelsea, New York, 1946.
[26] MacDuffee, C., *Vectors and Matrices*, Carus Mathematical Monograph Number 7, Math, Assoc. of Am., 1943.
[27] Marcus, M., *Basic Theorems in Matrix Theory*, National Bureau of Standards Applied Mathematics Series, No. 57, U.S. Government Printing Office, Washington, D.C., 1960.
[28] Marcus, M. and H. Minc, *A Survey of Matrix Theory and Matrix Inequalities*, Allyn and Bacon, Boston, 1964.
[29] Mathematics Division of National Physical Laboratory at Teddington, England, *Modern Computing Methods*, 2nd ed., Philosophical Library, New York, 1961.
[30] Meyer, H. A. (ed.), *Symposium on Monte Carlo Methods*, Wiley, New York, 1956.
[31] Morris, J., *The Escalator Method in Engineering Vibration Problems*, Wiley, New York, 1947.
[32] Perlis, S., *Theory of Matrices*, Addison-Wesley, Cambridge, Massachusetts, 1952.
[33] Ralston, A. and H. Wilf, *Mathematical Methods for Digital Computers*, Wiley, New York, 1960.
[34] Scarborough, J. B., *Numerical Mathematical Analysis*, Johns Hopkins, Baltimore, Maryland, 1958.
[35] Stiefel, E. L., P. Henrici, and H. Rutishauser, *Further Contributions to the Solution of Simultaneous Linear Equations and the Determination of Eigenvalues*, National Bureau of Standards Applied Mathematics Series No. 49, U.S. Government Printing Office, Washington, D.C., 1958.
[36] Taussky, O. and L. J. Paige (eds.), *Simultaneous Linear Equations and the Determination of Eigenvalues*, National Bureau of Standards Applied Mathematics Series No. 29, U.S. Government Printing Office, Washington, D.C., 1953.
[37] Taussky, O. (ed.), *Contributions to the Solution of Systems of Linear Equations and the Determination of Eigenvalues*, National Bureau of Standards Applied Mathematics Series No. 39, U.S. Government Printing Office, Washington, D.C., 1954.
[38] Todd, J. (ed.), *Survey of Numerical Analysis*, McGraw-Hill, New York, 1962.
[39] Turnbull, H. W., *The Theory of Determinants, Matrices and Invariants*, Blackie, Glasgow, 1929.
[40] Varga, R. S., *Matrix Iterative Analysis*, Prentice-Hall, Englewood Cliffs, New Jersey, 1962.
[41] Wilkinson, J. H., *Rounding Errors in Algebraic Processes*, Prentice-Hall, Englewood Cliffs, New Jersey, 1963.
[42] Wilkinson, J. H., *The Algebraic Eigenvalue Problem*, Clarendon Press, Oxford, 1965.

PAPERS

[43] Aitken, A. C., "On the Iterative Solution of a System of Linear Equations," *Proc. Roy. Soc. Edinburgh*, Sec. A, **63**, 52–60 (1950).
[44] Arms, R. J., L. D. Gates, and B. Zondek, "A Method of Block Iteration," *J. SIAM* **4**, 220–229 (1956).

[45] Bargmann, V., D. Montgomery, and J. von Neumann, "Solution of Linear Systems of High Order," *Institute for Advanced Study Report*, Princeton, New Jersey, Bur. Ord., Navy Dept., (1946).

[46] Bartlett, M. S., "An Inverse Matrix Adjustment Arising in Discriminant Analysis," *Ann. Math. Stat.* **22,** 107–111 (1951).

[47] Beckman, F. S. "The solution of Linear Equations by the Conjugate Gradient Method," in [33], pp. 62–72.

[48] Bellar, F. J., "An Iterative Solution of Large-Scale Systems of Simultaneous Linear Equations," *J. Soc. Indust. Appl. Math.* **9,** 189–193 (1961).

[49] Bingham, M. D., "A New Method for Obtaining the Inverse Matrix," *Jour. Amer. Stat. Assoc.* **35,** 530–534 (1941).

[50] Birkhoff, G., R. Varga, and D. M. Young, Jr., "Alternating Direction Implicit Methods," in [2], pp. 190–273.

[51] Busing, W. R. and Levy, H. A., "A Precedure for Inverting Large Symmetric Matrices," *Comm. ACM* **5,** 445–446 (1962).

[52] Caffrey, J., "Another Test Matrix for Determinants and Inverses," *Comm. ACM* **6,** 310 (1963).

[53] Cline, R. E., "A Class of Matrices to Test Inversion Procedures," *Comm. ACM* **7** 724–725 (1964).

[54] Crout, P. D., "A Short Method for Evaluating Determinants and Solving Systems of Linear Equations with Real or Complex Coefficients," *Amer. Inst. Elect. Eng.*, **60,** 1235–1240 (1941).

[55] Curtiss, J. H., "A Theoretical Comparison of the Efficiencies of Two Classical Methods and a Monte Carlo Method for Computing One Component of the Solution of a Set of Linear Algebraic Equations," in [30], pp. 191–233.

[56] Curtiss, J. H., "Monte Carlo Methods for the Iteration of Linear Operators," *J. Math. and Phys.*, **32,** 209–232 (1954).

[57] Fischbach, J. W., "Some Applications of Gradient Methods," in [9], pp. 59–72.

[58] Flanders, D. A. and G. Shortley, "Numerical Determination of Fundamental Modes," *J. Appl. Phys.*, **21,** 1326–1332 (1950).

[59] Forsythe, G. E., "Solving Linear Equations Can Be Interesting," *Bull. Amer. Math. Soc.* **59,** 299–329 (1953).

[60] Forsythe, G. E., "Theory of Selected Methods of Finite Matrix Inversion and Decomposition," Lectures given at UCLA, Summer 1951, 1954.

[61] Forsythe, G. E., "Tentative Classification of Methods and Bibliography on Solving Systems of Linear Equations," in [36], pp. 1–28.

[62] Forsythe, G. E. and R. A. Leibler, "Matrix Inversion by a Monte Carlo Method," *MTAC*, **4,** 127–129 (1950). Correction to the Article, "Matrix Inversion by a Monte Carlo Process," *MTAC*, **5,** 55 (1951).

[63] Fox, L., "Practical Methods for the Solution of Linear Equations and the Inversion of Matrices," in [37], pp. 1–54.

[64] Fox, L., "A Short Account of Relaxation Methods," *Quart. J. Mech. Appl. Math,.* **1,** 253–280 (1948).

[65] Fox, L., H. D. Huskey, and J. H. Wilkinson, "Notes on the Solution of Algebraic Linear Simultaneous Equations," *Quart. J. Mech. Appl. Math.* **1,** 149–173 (1948).

[66] Frank, W. L., "Solutions of Linear Systems by Richardson's Method,"J. *ACM*, **7,** 274–286 (1960).

[67] Geiringer, H., "On the Solution of Systems of Linear Equations by Certain Iteration Methods," *Reissner Anniversary Volume*, J. W. Edwards, Ann Arbor, Mich., 1949, pp. 365–393.

[68] Goldstine, H. H. and J. von Neumann, "Numerical Inverting of Matrices of High Order," *Bull. Amer. Math. Soc.* **53,** 1021–1099 (1947); and *Proc. Amer. Math. Soc.,* **2,** 188–202 (1951).
[69] Golub, G. H. and R. S. Varga, "Chebyshev Semi-Iterative Methods, Successive Overrelaxation Iterative Methods, and Second Order Richardson Iterative Methods," Part I: *Num. Math.* **3,** 147–156 (1961); Part II: *Num. Math.* **3,** 157–168 (1962).
[70] Halton, J. H., Sequential Monte Carlo, *Proc. Camb. Phil. Soc.* **58,** 57–78 (1962).
[71] Hansen, Eldon, "Interval Arithmetic in Matrix Computations, Part I," *J. SIAM Num. Anal. Ser. B* **2,** 308–320 (1965).
[72] Hayes, R. M., "Iterative Methods of Solving Linear Problems on Hilbert Space," in [37], pp. 71–103.
[73] Hestenes, M. R., "Inversion of Matrices by Biorthogonalization and Related Results," *J. Soc. Indust. Appl. Math.* **6,** 51–90 (1958).
[74] Hestenes, M. R., "The Conjugate Gradient Method for Solving Linear Systems," in [9], pp. 83–102.
[75] Hestenes, M. R. and E. Stiefel, "Method of Conjugate-Gradients for Solving Linear Systems," *J. Res. Natl. Bur. Std.* **49,** 409–436 (1952).
[76] Hotelling, H., "Some New Methods in Matrix Calculation," *Ann. Math. Stat.* **14,** 1–34 (1943).
[77] Hotelling, H., "Further Points on Matrix Calculation and Simultaneous Equations," *Ann. Math. Stat.* **14,** 440–441 (1943).
[78] Lanczos, C., "Iterative Solution of Large-Scale Systems," *J. Soc. Indust. Appl. Math.* **6,** 91–109 (1958).
[79] Lanczos, C., "Solution of Systems of Linear Equations by Minimized Iterations," *J. Res. Natl. Bur. Std.* **49,** 33–53 (1952).
[80] Lasor, W. S., "Test Matrix for Inversion," *Comm. ACM.* **6,** 102 (1963).
[81] Lietzke, M. H., R. W. Stoughton, and M. P. Lietzke, "A Comparison of Several Methods for Inverting Large Symmetric Positive Definite Matrices," *Math. Comput.* **18,** 449–456 (1964).
[82] Lotkin, M., A Set of Test Matrices," *MTAC,* **9,** 153–161 (1955).
[83] Newberry, A. R. C., Pei Matrix Eigenvectors," *Comm. ACM.* **9,** 515 (1963).
[84] Newman, M., "Matrix Computations," in [38], pp. 222–254.
[85] Newman, M. and J. Todd, "The Evaluation of Matrix Inversion Programs, "*J. Soc. Indust. Appl. Math.* **6,** 466–476 (1958).
[86] Orden, A., "Matrix Invertion and Related Topics by Direct Methods," in [33], pp. 39–55.
[87] Oswald, F. J., "Matrix Inversion by Monte Carlo Methods," in [33], pp. 78–83.
[88] Ortega, J. M., "Generation of Test Matrices by Similarity Transformations," *Comm, ACM.* **7,** 377–378 (1964).
[89] Pei, M. L., "A Test Matrix for Inversion Procedures," *Comm. ACM.* **5,** 508 (1962).
[90] Rodman, R. D., "A Note on a Set of Test Matrices for Inversion," *Comm. ACM.* **6,** 515 (1963).
[91] Rushton, S., "On Least-Squares Fitting by Orthonormal Polynomials Using the Choleski Method," *J. Res. Stat. Soc.* (*B*) **13,** 92–99 (1951).
[92] Savage, R. and E. Lukacs, "Tables of Finite Segments of the Hilbert Matrix," in [37], pp. 105–108.
[93] Schechter, S., "Quasi-Tridiagonal Matrices and Typr-Insensitive Difference Equations," AEC Computing and Applied Mathematics Center, Institute of Mathematical Sciences, NYU, NYO-2542, May 1, 1959.

[94] Sheldon, J. W., "Iterative Methods for the Solution of Elliptic Partial Differential Equations," in [33], pp. 144–156.
[95] Sherman, J., "Computations Relating to Inverse Matrices," in [36], pp. 123–124.
[96] Sherman, J. and W. J. Morrsion, "Adjustment of an Inverse Matrix Corresponding to Changes in a Given Column or a Given Row of the Original Matrix," *Ann. Math. Stat.* **21**, 124 (1949).
[97] Stein, M. L., "Gradient Methods in the Solution of Systems of Linear Equations," *J. Res. Natl. Bur. Std.* **48**, 407–413 (1952).
[98] Taussky, O., "Note on the Condition of Matrices," *MTAC* **4**, 111–112 (1950).
[99] Todd, J., "The Condition of a Certain Matrix," *Proc. Camb. Phil. Soc.* **46**, 116(1950).
[100] Todd, J., "The Condition of the Finite Segments of the Hilbert Matrix," in [37], pp. 109–116.
[101] Turing, A. M., "Rounding-Off Errors in Matrix Processes," *Quart. J. Mech. Appl. Math.* **1**, 287–308 (1948).
[102] Van Norton, R., "The Solution of Linear Equations by the Gauss-Seidel Method," in [33], pp. 56–61.
[103] Varga, R. S., A Comparison of the Successive Overrelaxation Method and Semi-Iterative Methods Using Chebyshev Polynomials," *J. SIAM* **5**, 39–46 (1957).
[104] Wachspress, E. L., "Optimum Alternating- Direction-Implicit Iteration Parameters for a Model Problem," *J. SIAM* **10**, 339–350 (1962).
[105] Wasow, W. R., "A Note on the Inversion of Matrices by Random Walks," *MTAC*, **6**, 78–81 (1952).
[106] Wilf, H. S., "Matrix Inversion by the Annihilation of Rank, *J. Soc. Indust. Appl. Math.* **7**, 149–151 (1959).
[107] Wilf, H. S., "Matrix Inversion by the Method of Rank Annihilation," in [33], pp. 73–77.
[108] Wilkinson, J. H., "Error Analysis of Direct Methods of Matrix Inversion," *J. ACM* **8**, 281–330 (1961).
[109] Young, D., "On the Solution of Linear Systems by Iteration," in [9], pp. 283–298.
[110] Young, D., "On Richardson's Method for Solving Linear Systems with Positive Definite Matrices," *J. Math. and Phys.* **32**, 243–255 (1953).
[111] Young, D., "Iterative Methods for Solving Partial Differential Equations of the Elliptic Type," *Trans. Amer. Math. Soc.* **76**, 92–111 (1954).
[112] Young, D., "The Numerical Solution of Elliptic and Parabolic Partial Differential Equations," in [4], pp. 373–419.
[113] Young. D., "The Numerical Solution of Elliptic and Parabolic Partial Differential Equations," in [38], pp. 380–438.
[114] Young, D., and L. V. Ehrlich, "Some Numerical Studies of Iterative Methods for Solving Elliptic Difference Equations," in [24], pp. 143–162.

ADDITIONAL REFERENCES FOR THE EIGENVALUES OF TEST MATRICES CONTAINED IN APPENDIX C:

[115] Bodewig, E., "A Practical Refutation of the Iteration Method for the Eigenproblem," *MTAC* **8**, 237–240 (1954).
[116] Eberlein, P. J., "A Jacobi-Like Method for the Automatic Computation of Eigenvalues and Eigenvectors of an Arbitrary Matrix," *J. SIAM* **10**, 74–88 (1962).
[117] Eberlein, P. J., "A Two-Parameter Test Matrix," *Math. Comput.* **18**, 296–298 (1964).

[118] Frank, W. L., "Computing Eigenvalues of Complex Matrices by Determinant Evaluation and by Methods of Danilewski and Wielandt," *J. SIAM* **6,** 378–392 (1958).

[119] Parlett, B., "Laguerre's Method Applied to the Matrix Eigenvalue Problem," *Math. Compute.* **18,** 464–485 (1964).

[120] Rosser, J. B., C. Lanczos, M. R. Hestenes, and W. Karush, "Separation of Close Eigenvalues of a Real Symmetric Matrix," *J. Res. Natl. Bur. Std.* **47,** 291–297 (1951).

[121] Taussky, O., and M. Marcus, "Eigenvalues of Finite Matrices," in [38], pp. 279–313.

[122] White, P. A., "The Computation of Eigenvalues and Eigenvectors of a Matrix," *J. Soc. Indust. Appl. Math.* **6,** 393–437 (1958).

[123] Wilkinson, J. H., The Use of Iterative Methods for Finding the Latent Roots and Vectors of Matrices," *MTAC* **9,** 184–191 (1955).

[124] Wilkinson, J. H., "The Evaluation of the Zeros of Ill–Conditioned Polynomials," Parts I and II, *Num. Math.* **1,** 150–180 (1959).

[125] Wilkinson, J. H., "Error Analysis of Floating-Point Computation," *Num. Math.* **2,** 319–340 (1960).

[126] Wilkinson, J. H., "Rigorous Error Bounds for Computed Eigensystems," *Comput. J.* **4,** 230–241 (1961).

SYMBOL TABLE

A	an $n \times n$ matrix
$AX = B$	r linear systems where B is an $n \times r$ matrix
$Ax = b$	one linear system
I	the $n \times n$ identity matrix
λ	eigenvalue
L	lower triangular matrix
U	upper triangular matrix
A^{-1}	inverse of A
A', x'	transpose of matrix A, transpose of vector x
A^*	conjugate of A
$\|A\|$, $d(A)$, $\det(A)$	determinant of A
D	diagonal matrix
v	eigenvector
x, y	vectors
δ_{ij}	Kronecker delta is 1 if $i = j$, 0 otherwise
H	Hermitian matrix
$\\|A\\|$	norm of matrix A
$\\|x\\|$	norm of vector x
R	orthogonal matrix
V	unitary matrix
$\bar{a}$	conjugate of number a
ϵ_i	ith unit vector
(x, y), $x \cdot y$, $x'y$	scalar product of two vectors
r, η	residual vector
e	error vector

$\binom{N}{R}$	binomial coefficient
$\lvert a \rvert$	absolute value of the number a
$x^{(k)}$	kth iterate of the vector x
$x_i^{(k)}$	ith component of the kth iterate of the vector x
$A^{(k)}$	kth iterate of the matrix A
a_{ij}	element in the ith row and jth column of the matrix $A = (a_{ij})$
α, β	scalars
A_{ij}	cofactor of matrix element a_{ij}
$a_{ij}^{(k)}$	element in ith row, jth column of kth iterate of matrix A
$T, T^{(k)}$	iteration matrix
$\lambda(T)$	spectral radius of T
$R(T)$	rate of convergence of the iteration whose iteration matrix is T
h	solution vector of $Ax = b$
ω	relaxation or acceleration parameter
SOR	successive overrelaxation
ADI	alternating direction implicit
Tr A	trace of matrix A

INDEX